Python Natural Language Processing Cookbook

Over 60 recipes for building powerful NLP solutions using Python and LLM libraries

Zhenya Antić, Ph.D.

Saurabh Chakravarty, Ph.D.

Python Natural Language Processing Cookbook

Group Product Manager: Niranjan Naikwadi

Publishing Product Manager: Tejashwini R

Book Project Manager: Aparna Nair

Lead Editor: Sushma Reddy

Senior Editor: Paridhi Agarwal

Technical Editor: Rahul Limbachiya

Copy Editor: Safis Editing

Proofreader: Sushma Reddy

Indexer: Rekha Nair

Production Designer: Gokul Raj S T

DevRel Marketing Coordinator: Nivedita Singh

First published: March 2021

Second edition: September 2024

Production reference: 2200525

Published by Packt Publishing Ltd.

Grosvenor House

11 St Paul's Square

Birmingham

B3 1RB, UK

ISBN 978-1-80324-574-4

www.packtpub.com

I would like to thank those who helped bring this book to life. The whole Packt team, including Tejashwini, Aparna Nair, Sushma Reddy, Paridhi Agarwal, and Tushar Gupta, has been really helpful, providing insights and tips when necessary. I would also like to thank John Ortega, the technical reviewer, for his extensive and insightful comments about the code. A heartfelt thank you goes to Saurabh, who agreed to come on this journey with me.

Finally, I would like to thank my family for their constant support.

Zhenya Antić

This book is dedicated to the memory of my late father, Mr. Dilip Kumar Chakravarty, and father-in-law, Mr. Sanjiv Kumar Bose. You are both deeply missed. I also owe a debt of gratitude to Aaliya, Ananya, Anima, Arnabh, Arpita, Dipankar, Nilima, Pratima, Purnima, Rekha, Shikha, Somnath, Shubha, Sushma, Tapan, Tina, and Vladimir. You have all significantly influenced my life and supported me through both good and bad times. Special thanks to my doctoral advisor, Dr. Edward Fox, for teaching me everything I know about NLP and AI. Finally, I want to thank the editing team at Packt Publishing who were patient with me throughout the writing process.

Saurabh Chakravarty

Foreword

I am excited to see the high quality of this practical, timely, and comprehensive contribution to the field of **natural language processing (NLP)** by such a knowledgeable team of writers. This book will clearly be of great benefit to practitioners.

Having taught related courses for over four decades, I also heartily recommend this work to the many students interested in Python, text processing, text analytics/mining, information retrieval, digital libraries, and related fields.

Given the rapid advances in NLP—particularly in infrequently covered areas like topic modeling (Chapter 6), the importance of human-in-the-loop approaches such as with visualization (Chapter 7), the numerous benefits of transformers (Chapter 8), and the wide-ranging applications of GenAI and LLMs (Chapter 10), including summarization and other aspects of natural language understanding (Chapter 9)—readers will find this book current, up-to-date, and highly relevant.

The many examples and code recipes are especially helpful for those who, like my students, learn best by just-in-time or problem-based learning.

Edward A. Fox, Ph.D.
Emeritus Professor of Computer Science, Virginia Tech

Contributors

About the authors

Zhenya Antić, Ph.D. is an expert in AI and NLP. She is currently the Director of AI Automation at Arch Insurance, where she leads initiatives in Intelligent Document Processing and applies various AI solutions to complex problems. With extensive consulting experience, Zhenya has worked on numerous NLP projects with various companies. She holds a Ph.D. in Linguistics from the University of California, Berkeley, and a B.S. in Computer Science from the Massachusetts Institute of Technology.

Saurabh Chakravarty, Ph.D. is a seasoned veteran in the software industry with over 20 years of experience in software development. A software developer at heart, he is passionate about programming. He has held various roles, including architect, lead engineer, and software developer, specializing in AI and large-scale distributed systems. Saurabh has worked with Microsoft, Rackspace, and Accenture, as well as with a few startups. He holds a Ph.D. in Computer Science with a specialization in NLP from Virginia Tech, USA. Saurabh lives in California with his wife, Tina, and daughter, Aaliya, and works for AWS in Santa Clara, California.

Disclaimer: The opinions expressed in this book are entirely my own and do not reflect those of my employer, Amazon Web Services (AWS). These views have not been expressed in my capacity as an AWS employee.

I would also like to express my gratitude to the reviewers from the AWS team for dedicating their personal time to review the content.

About the reviewer

John E. Ortega is a research scientist and professor of natural language processing at Columbia University. He holds a Ph.D. in Computer Science from the Universitat d'Alacant in Spain and has completed postgraduate research at the University of Santiago de Compostela and Northeastern University. In addition to founding two companies, List Properties and Vidpal, he has worked as a consultant in the United States and Europe for the past 21 years. Ortega has also served as a senior NLP research scientist at AIG and Nuance Communications, and as an Executive Director for AI and Machine Learning at JPMorgan Chase. His volunteer work includes organizing kids' basketball tournaments, maintaining park grounds, and rescuing elephants.

Table of Contents

3

Representing Text – Capturing Semantics   49

4

Classifying Texts   81

5

Getting Started with Information Extraction   111

6

Topic Modeling   143

7

Visualizing Text Data   169

8

Transformers and Their Applications   193

9

Natural Language Understanding   211

Preface

Python is the most widely used language for **natural language processing** (**NLP**) thanks to its extensive tools and libraries for analyzing text and extracting computer-usable data. This book will take you through a range of techniques for text processing, from basics such as parsing parts of speech to complex topics such as topic modeling, text classification, and visualization.

Starting with an overview of NLP, the book presents recipes for dividing text into sentences, stemming and lemmatization, removing stopwords, and parts-of-speech tagging to help you to prepare your data. You will then learn about ways of extracting and representing grammatical information, such as dependency parsing and anaphora resolution; discover different ways of representing the semantics using bag of words, TF-IDF, word embeddings, and BERT; and develop skills for text classification using keywords, SVMs, LSTMs, and other techniques.

As you advance, you will also see how to extract information from text, implement unsupervised and supervised techniques for topic modeling, and perform topic modeling of short texts, such as tweets. Additionally, the book covers visualizations of text data.

Finally, this book introduces Transformer-based models and how to utilize them to perform another set of novel NLP tasks. These encoder-decoder-based models are deep neural-network-based models and have been trained on large text corpora. These models have performed or exceeded the state of the art on various NLP tasks. Especially novel are the decoder-based generative models, which have the capability to generate text based on the context provided to them. Some of these models have reasoning capabilities built into them. These models will take NLP into the next era and make it a part of mainstream technology applications.

By the end of this NLP book, you will have developed the skills to use a powerful set of tools for text processing.

Who this book is for

Data scientists, machine learning engineers, and developers familiar with basic programming and data science concepts can gain practical insights from this book. It serves as a primer to introduce the concepts of NLP and their practical applications.

The roles that are the target of this book are the following:

Data scientists: As a data scientist, you will gain an understanding of how to work with text. Intermediate knowledge of Python will help you to get the most out of this book. If you are already an NLP practitioner, this book will serve as a code reference when working on your projects.

Software engineers and architects: Developers who want to build capability in the domain of NLP will be introduced to all the fundamental and advanced uses of NLP in text processing. This will help you level up on your knowledge and build yourself up to develop solutions using NLP when required.

Product managers: Though this book contains code examples for the recipes, each of these recipes is accompanied by explanations of why certain steps are being performed and what the end output for those steps is. This makes it a useful resource for product managers who want to understand what is possible with a certain NLP recipe, which would enable them to envision novel solutions using it.

What this book covers

Chapter 1, Learning NLP Basics, introduces the very basics of NLP. The recipes in this chapter show the basic preprocessing steps that are required for further NLP work. We show how to tokenize text, or divide it into sentences and words; assign parts of speech to individual words; lemmatize them, or get their canonical forms; and remove stopwords.

Chapter 2, Playing with Grammar, shows how to get grammatical information from text. This information could be useful in determining relationships between different entities mentioned in the text. We start by showing how to determine whether a noun is singular or plural. We then show how to get a dependency parse that shows relationships between words in a sentence. Then, we demonstrate how to get noun chunks, or nouns with their dependent words, such as adjectives. After that, we look at parsing out the subjects and objects of a sentence. Finally, we show how to use a regular-expression-style matcher to extract grammatical phrases in a sentence.

Chapter 3, Representing Text – Capturing Semantics, looks at different ways of representing text for further processing in NLP models. Since computers cannot deal with words directly, we need to encode them in vector form. In order to demonstrate the effectiveness of different methods of encoding, we first create a simple classifier and then use it with different encoding methods. We look at the following encoding methods: bag-of-words, N-gram model, TF-IDF, word embeddings, BERT, and OpenAI embeddings. We also show how to train your own bag-of-words model and demonstrate how to create a simple **retrieval-augmented generation (RAG)** solution.

Chapter 4, Classifying Texts, shows various ways of carrying out text classification, one of the most common NLP tasks. First, we show how to preprocess the dataset in order to prepare it for classification. Then, we demonstrate different classifiers, including a rule-based classifier, an unsupervised classifier via K-means, training an SVM for classification, training a spaCy model for text classification, and, finally, using OpenAI GPT models to classify texts.

Chapter 5, Getting Started with Information Extraction, shows how to extract information from text, another very important NLP task. We start off with using regular expressions for simple information extraction. We then look at how to use the Levenshtein distance to handle misspellings. Then, we show how to extract characteristic keywords from different texts. We look at how to extract named entities using spaCy, and how to train your own custom spaCy NER model. Finally, we show how to fine-tune a BERT NER model.

Chapter 6, Topic Modeling, shows how to determine topics of text using various unsupervised methods, including LDA, community detection with BERT embeddings, K-means clustering, and BERTopic. Finally, we use contextualized topic models that work with multilingual models and inputs.

Chapter 7, Visualizing Text Data, focuses on using various tools to create informative visualizations of text data and processing. We create graphic representations of the dependency parse, parts of speech, and named entities. We also create a confusion matrix plot and word clouds. Finally, we use pyLDAvis and BERTopic to visualize topics in a text.

Chapter 8, Transformers and Their Applications, provides an introduction to Transformers. This chapter begins by demonstrating how to transform text into a format suitable for internal processing by a Transformer model. It then explores techniques for text classification using pre-trained Transformer models. Additionally, the chapter delves into text generation with Transformers, explaining how to tweak the generation parameters to produce coherent and natural-sounding text. Finally, it covers the application of Transformers in language translation.

Chapter 9, Natural Language Understanding, covers NLP techniques that help infer the information contained in a piece of text. This chapter begins with a discussion on question-answering in both open and closed domains, followed by methods for answering questions from document sources using extractive and abstractive approaches. Subsequent sections cover text summarization and sentence entailment. The chapter concludes with explainability techniques, which demonstrate how models make classification decisions and how different parts of the text contribute to the assigned class labels.

Chapter 10, Generative AI and Large Language Models, introduces open source **Large Language Models (LLMs)** such as Mistral and Llama, demonstrating how to use prompts to generate text based on simple human-defined requirements. It further explores techniques for generating Python code and SQL statements from natural language instructions. Finally, it presents methods for utilizing a sophisticated closed source LLM from OpenAI to orchestrate custom task agents. These agents collaborate to answer complex questions requiring web searches and basic arithmetic to arrive at an end solution.

To get the most out of this book

You will need an understanding of the Python programming language and how to manage and install packages for it. Knowledge of Jupyter Notebook would be useful, though it is not required. For package management, the knowledge of `poetry` package management is recommended, though you can make the examples work via `pip` too. For recipes to be able to use GPUs (if present) in the system, ensure that the latest GPU device drivers are installed along with the CUDA/cuDNN dependencies.

Software/hardware covered in the book	Operating system requirements
Python 3.10	Windows, macOS, or Linux
Poetry	Windows, macOS, or Linux
Jupyter Notebook (optional)	Windows, macOS, or Linux

If you are using the digital version of this book, we advise you to type the code yourself or access the code from the book's GitHub repository (a link is available in the next section). Doing so will help you avoid any potential errors related to the copying and pasting of code.

Download the example code files

You can download the example code files for this book from GitHub at https://github.com/PacktPublishing/Python-Natural-Language-Processing-Cookbook-Second-Edition. If there is an update to the code, it will be updated in the GitHub repository.

We also have other code bundles from our rich catalog of books and videos available at https://github.com/PacktPublishing/. Check them out!

Conventions used

There are several text conventions used throughout this book.

Code in text: Indicates code words in text, database table names, folder names, filenames, file extensions, pathnames, dummy URLs, user input, and *Twitter/X* handles. Here is an example: "Instantiate a tokenizer of the bert-base-cased type."

A block of code is set as follows:

```
from transformers import pipeline
import torch
```

When we wish to draw your attention to a particular part of a code block, the relevant lines or items are set in bold:

```
from transformers import T5Tokenizer, T5ForConditionalGeneration
import torch
device = torch.device("cuda" if torch.cuda.is_available() else "cpu")
```

Any command-line input or output is written as follows:

```
Classification Results
{'accuracy': 0.88,
'precision': 0.92,
'recall': 0.84}
```

Bold: Indicates a new term, an important word, or words that you see onscreen. For instance, words in menus or dialog boxes appear in **bold**. Here is an example: "Select **System info** from the **Administration** panel."

> **Tips or important notes**
> Appear like this.

Get in touch

Feedback from our readers is always welcome.

General feedback: If you have questions about any aspect of this book, email us at `customercare@packtpub.com` and mention the book title in the subject of your message.

Errata: Although we have taken every care to ensure the accuracy of our content, mistakes do happen. If you have found a mistake in this book, we would be grateful if you would report this to us. Please visit `www.packtpub.com/support/errata` and fill in the form.

Piracy: If you come across any illegal copies of our works in any form on the internet, we would be grateful if you would provide us with the location address or website name. Please contact us at `copyright@packt.com` with a link to the material.

If you are interested in becoming an author: If there is a topic that you have expertise in and you are interested in either writing or contributing to a book, please visit `authors.packtpub.com`.

Share Your Thoughts

Once you've read *Python Natural Language Processing Cookbook*, we'd love to hear your thoughts! Scan the QR code below to go straight to the Amazon review page for this book and share your feedback.

`https://packt.link/r/1-803-24574-3`

Your review is important to us and the tech community and will help us make sure we're delivering excellent quality content.

Download a free PDF copy of this book

Thanks for purchasing this book!

Do you like to read on the go but are unable to carry your print books everywhere?

Is your eBook purchase not compatible with the device of your choice?

Don't worry, now with every Packt book you get a DRM-free PDF version of that book at no cost.

Read anywhere, any place, on any device. Search, copy, and paste code from your favorite technical books directly into your application.

The perks don't stop there, you can get exclusive access to discounts, newsletters, and great free content in your inbox daily

Follow these simple steps to get the benefits:

1. Scan the QR code or visit the link below

https://packt.link/free-ebook/978-1-80324-574-4

2. Submit your proof of purchase
3. That's it! We'll send your free PDF and other benefits to your email directly

1

Learning NLP Basics

While working on this book, we were focusing on including recipes that should be useful for a wide variety of NLP projects. They range from simple to advanced, from dealing with grammar to dealing with visualizations, and in many of them, options for languages other than English are included. In this new edition, we have included new topics that cover using GPT and other large language models, explainable AI, a new chapter on transformers, and natural language understanding. We hope you find the book useful.

The format of the book is that of a *programming cookbook*, where each recipe is a short mini-project with a concrete goal and a sequence of steps that need to be performed. There are few theoretical explanations and a focus on the practical goals and what needs to be done to achieve them.

Before we can get on with the real work of NLP, we need to prepare our text for processing. This chapter will show you how to do it. By the end of the chapter, you will be able to have a list of words in a text with their parts of speech and lemmas or stems, and with very frequent words removed.

Natural Language Toolkit (**NLTK**) and **spaCy** are two important packages that we will be working with in this chapter and throughout the book. Some other packages we will be using in the book are PyTorch and Hugging Face Transformers. We will also utilize the OpenAI API with the GPT models.

The recipes included in this chapter are as follows:

- Dividing text into sentences
- Dividing sentences into words – tokenization
- Part of speech tagging
- Combining similar words – lemmatization
- Removing stopwords

Technical requirements

Throughout this book, we will use **Poetry** to manage the Python package installations. You can use the latest version of Poetry since it conserves the previous versions' functionality. Once you install

Poetry, managing which packages to install will be very easy. We will be using **Python 3.9** throughout the book. You will also need to have **Jupyter** installed in order to run the notebooks.

> **Note**
> You may try to use Google Colab in order to run the notebooks but you will need to tweak the code to make it work with Colab.

Follow these installation steps:

1. Install **Git**: `https://github.com/git-guides/install-git`.

2. Install **Poetry**: `https://python-poetry.org/docs/#installation`.

3. Install **Jupyter**: `https://jupyter.org/install`.

4. Clone the GitHub repository that contains all the code from this book (`https://github.com/PacktPublishing/Python-Natural-Language-Processing-Cookbook-Second-Edition`) by issuing the following command in the terminal:

```
git clone https://github.com/PacktPublishing/Python-Natural-Language-
Processing-Cookbook-Second-Edition.git
```

5. In the directory that contains the `pyproject.toml` file, run the commands using the terminal:

```
poetry install
poetry shell
```

6. Start the notebook engine:

```
jupyter notebook
```

Now, you should be able to run all the notebooks in your cloned repository.

If you prefer not to use Poetry, you can set up a virtual environment using the `requirements.txt` file provided in the book repository. You can do this in one of two ways. You can use `pip`:

```
pip install -r requirements.txt
```

You can also use `conda`:

```
conda create --name <env_name> --file requirements.txt
```

Dividing text into sentences

When we work with text, we can work with text units on different scales: the document itself, such as a newspaper article, the paragraph, the sentence, or the word. Sentences are the main unit of processing in many NLP tasks. For example, when we send data over to **Large Language Models** (**LLMs**), we frequently want to add some context to the prompt. In some cases, we would like that context to include sentences from a text so that the model can extract some important information from that text. In this section, we will show you how to divide a text into sentences.

Getting ready

For this part, we will be using the text of the book *The Adventures of Sherlock Holmes*. You can find the whole text in the book's GitHub file (`https://github.com/PacktPublishing/Python-Natural-Language-Processing-Cookbook-Second-Edition/blob/main/data/sherlock_holmes.txt`). For this recipe we will need just the beginning of the book, which can be found in the file at `https://github.com/PacktPublishing/Python-Natural-Language-Processing-Cookbook-Second-Edition/blob/main/data/sherlock_holmes_1.txt`.

In order to do this task, you will need the NLTK package and its sentence tokenizers, which are part of the Poetry file. Directions to install Poetry are described in the *Technical requirements* section.

How to do it...

We will now divide the text of a small piece of *The Adventures of Sherlock Holmes*, outputting a list of sentences. (Reference notebook: `https://github.com/PacktPublishing/Python-Natural-Language-Processing-Cookbook-Second-Edition/blob/main/Chapter01/dividing_text_into_sentences_1.1.ipynb`.) Here, we assume that you are running the notebook, so the paths are all relative to the notebook location:

1. Import the file utility functions from the `util` folder (`https://github.com/PacktPublishing/Python-Natural-Language-Processing-Cookbook-Second-Edition/blob/main/util/file_utils.ipynb`):

   ```
   %run -i "../util/file_utils.ipynb"
   ```

2. Read in the book part text:

   ```
   sherlock_holmes_part_of_text = read_text_file("../data/sherlock_holmes_1.txt")
   ```

   The `read_text_file` function is located in the `util` notebook we imported previously. Here is its source code:

   ```
   def read_text_file(filename):
       file = open(filename, "r", encoding="utf-8")
       return file.read()
   ```

3. Print out the resulting text to make sure everything worked correctly and the file loaded:

```
print(sherlock_holmes_part_of_text)
```

The beginning of the printout will look like this:

```
To Sherlock Holmes she is always _the_ woman. I have seldom
heard him
mention her under any other name. In his eyes she eclipses and
predominates the whole of her sex…
```

4. Import the `nltk` package:

```
import nltk
```

5. If this is the first time you are running the code, you will need to download tokenizer data. You will not need to run this command after that:

```
nltk.download('punkt')
```

6. Initialize the tokenizer:

```
tokenizer = nltk.data.load("tokenizers/punkt/english.pickle")
```

7. Divide the text into sentences using the tokenizer. The result will be a list of sentences:

```
sentences_nltk = tokenizer.tokenize(
    sherlock_holmes_part_of_text)
```

8. Print the result:

```
print(sentences_nltk)
```

It should look like this. There are newlines inside the sentences that come from the book formatting. They are not necessarily sentence endings:

```
['To Sherlock Holmes she is always _the_ woman.', 'I have seldom
heard him\nmention her under any other name.', 'In his eyes she
eclipses and\npredominates the whole of her sex.', 'It was not
that he felt any emotion\nakin to love for Irene Adler.', 'All
emotions, and that one particularly,\nwere abhorrent to his
cold, precise but admirably balanced mind.', 'He\nwas, I take
it, the most perfect reasoning and observing machine that\nthe
world has seen, but as a lover he would have placed himself in
a\nfalse position.', 'He never spoke of the softer passions,
save with a gibe\nand a sneer.', 'They were admirable things
for the observer—excellent for\ndrawing the veil from men's
motives and actions.', 'But for the trained\nreasoner to admit
such intrusions into his own delicate and finely\nadjusted
temperament was to introduce a distracting factor which might\
nthrow a doubt upon all his mental results.', 'Grit in a
sensitive\ninstrument, or a crack in one of his own high-power
lenses, would not\nbe more disturbing than a strong emotion in
```

```
a nature such as his.', 'And\nyet there was but one woman to
him, and that woman was the late Irene\nAdler, of dubious and
questionable memory.']
```

9. Print the number of sentences in the result; there should be 11 sentences in total:

    ```
    print(len(sentences_nltk))
    ```

    This gives the result:

    ```
    11
    ```

Although it might seem straightforward to divide a text into sentences by just using a regular expression to split it at the periods, in reality, it is more complicated. We use periods in places other than ends of sentences; for example, after abbreviations – for example, "Dr. Smith will see you now." Similarly, while all sentences in English start with a capital letter, we also use capital letters for proper names. The approach used in `nltk` takes all these points into consideration; it is an implementation of an unsupervised algorithm presented in `https://aclanthology.org/J06-4003.pdf`.

There's more...

We can also use a different strategy to parse the text into sentences, employing the other very popular NLP package, **spaCy**. Here is how it works:

1. Import the spaCy package:

    ```
    import spacy
    ```

2. The first time you run the notebook, you will need to download a spaCy model. The model is trained on a large amount of English text and there are several tools that can be used with it, including the sentence tokenizer. Here, I'm downloading the smallest model, but you might try other ones (see `https://spacy.io/usage/models/`):

    ```
    !python -m spacy download en_core_web_sm
    ```

3. Initialize the spaCy engine:

    ```
    nlp = spacy.load("en_core_web_sm")
    ```

4. Process the text using the spaCy engine. This line assumes that you have the `sherlock_holmes_part_of_text` variable initialized. If not, you need to run one of the earlier cells where the text is read into this variable:

    ```
    doc = nlp(sherlock_holmes_part_of_text)
    ```

5. Get the sentences from the processed `doc` object, and print the resulting array and its length:

    ```
    sentences_spacy = [sentence.text for sentence in doc.sents]
    print(sentences_spacy)
    print(len(sentences_spacy))
    ```

The result will look like this:

```
['To Sherlock Holmes she is always _the_ woman.', 'I have seldom
heard him\nmention her under any other name.', 'In his eyes she
eclipses and\npredominates the whole of her sex.', 'It was not
that he felt any emotion\nakin to love for Irene Adler.', 'All
emotions, and that one particularly,\nwere abhorrent to his
cold, precise but admirably balanced mind.', 'He\nwas, I take
it, the most perfect reasoning and observing machine that\nthe
world has seen, but as a lover he would have placed himself in
a\nfalse position.', 'He never spoke of the softer passions,
save with a gibe\nand a sneer.', 'They were admirable things
for the observer—excellent for\ndrawing the veil from men's
motives and actions.', 'But for the trained\nreasoner to admit
such intrusions into his own delicate and finely\nadjusted
temperament was to introduce a distracting factor which might\
nthrow a doubt upon all his mental results.', 'Grit in a
sensitive\ninstrument, or a crack in one of his own high-power
lenses, would not\nbe more disturbing than a strong emotion in
a nature such as his.', 'And\nyet there was but one woman to
him, and that woman was the late Irene\nAdler, of dubious and
questionable memory.']
11
```

An important difference between spaCy and NLTK is the time it takes to complete the sentence-splitting process. The reason for this is that spaCy loads a language model and uses several tools in addition to the tokenizer, while the NLTK tokenizer has only one function: to separate the text into sentences. We can time the execution by using the `time` package and putting the code to split the sentences into the `main` function:

```
import time

def split_into_sentences_nltk(text):
    sentences = tokenizer.tokenize(text)
    return sentences

def split_into_sentences_spacy(text):
    doc = nlp(text)
    sentences = [sentence.text for sentence in doc.sents]
    return sentences

start = time.time()
split_into_sentences_nltk(sherlock_holmes_part_of_text)
print(f"NLTK: {time.time() - start} s")

start = time.time()
split_into_sentences_spacy(sherlock_holmes_part_of_text)
print(f"spaCy: {time.time() - start} s")
```

The spaCy algorithm takes 0.019 seconds, while the NLTK algorithm takes 0.0002. The time is calculated by subtracting the current time (`time.time()`) from the start time that is set at the beginning of the code block. It is possible that you will get slightly different values.

The reason why you might use spaCy is if you are doing other processing with the package along with splitting it into sentences. The spaCy processor does many other things, and that is why it takes longer. If you are using other features of spaCy, there is no reason to use NLTK just for sentence splitting, and it's better to employ spaCy for the whole pipeline.

It is also possible to use only the tokenizer without other tools from spaCy. Please see their documentation for more information: `https://spacy.io/usage/processing-pipelines`.

> **Important note**
> spaCy might be slower, but it is doing many more things in the background, and if you are using its other features, use it for sentence splitting as well.

See also

You can use NLTK and spaCy to divide texts in languages other than English. NLTK includes tokenizer models for Czech, Danish, Dutch, Estonian, Finnish, French, German, Greek, Italian, Norwegian, Polish, Portuguese, Slovene, Spanish, Swedish, and Turkish. In order to load those models, use the name of the language followed by the `.pickle` extension:

```
tokenizer = nltk.data.load("tokenizers/punkt/spanish.pickle")
```

See the NLTK documentation to find out more: `https://www.nltk.org/index.html`.

Likewise, spaCy has models for other languages: Chinese, Dutch, English, French, German, Greek, Italian, Japanese, Portuguese, Romanian, Spanish, and others. These models are trained on text in those languages. In order to use those models, you would have to download them separately. For example, for Spanish, use this command to download the model:

```
python -m spacy download es_core_news_sm
```

Then, put this line in the code to use it:

```
nlp = spacy.load("es_core_news_sm")
```

See the spaCy documentation to find out more: `https://spacy.io/usage/models`.

Dividing sentences into words – tokenization

In many instances, we rely on individual words when we do NLP tasks. This happens, for example, when we build semantic models of texts by relying on the semantics – of individual words, or when we are looking for words with a specific part of speech. To divide text into words, we can use NLTK and spaCy to do this task for us.

Getting ready

For this part, we will be using the same text of the book *The Adventures of Sherlock Holmes*. You can find the whole text in the book's GitHub repository. For this recipe, we will need just the beginning of the book, which can be found in the `sherlock_holmes_1.txt` file.

In order to do this task, you will need the NLTK and spaCy packages, which are part of the Poetry file. Directions to install Poetry are described in the *Technical requirements* section.

(Notebook reference: `https://github.com/PacktPublishing/Python-Natural-Language-Processing-Cookbook-Second-Edition/blob/main/Chapter01/dividing_sentences_into_words_1.2.ipynb`.)

How to do it

The process is as follows:

1. Import the `file_utils` notebook. Effectively, we run the `file_utils` notebook inside this one so we have access to its defined functions and variables:

    ```
    %run -i "../util/file_utils.ipynb"
    ```

2. Read in the book snippet text:

    ```
    sherlock_holmes_part_of_text = read_text_file("../data/sherlock_
    holmes_1.txt")
    print(sherlock_holmes_part_of_text)
    ```

    The result should look like this:

    ```
    To Sherlock Holmes she is always _the_ woman. I have seldom
    heard him
    mention her under any other name. In his eyes she eclipses and
    predominates the whole of her sex... [Output truncated]
    ```

3. Import the `nltk` package:

    ```
    import nltk
    ```

4. Divide the input into words. Here, we use the NLTK word tokenizer to split the text into individual words. The output of the function is a Python list of the words:

    ```
    words = nltk.tokenize.word_tokenize(
        sherlock_holmes_part_of_text)
    print(words)
    print(len(words))
    ```

5. The output will be the list of words in the text and the length of the words list:

```
['To', 'Sherlock', 'Holmes', 'she', 'is', 'always', '_the_',
'woman', '.', 'I', 'have', 'seldom', 'heard', 'him', 'mention',
'her', 'under', 'any', 'other', 'name', '.', 'In', 'his',
'eyes', 'she', 'eclipses', 'and', 'predominates', 'the',
'whole', 'of', 'her', 'sex', '.', 'It', 'was', 'not', 'that',
'he', 'felt', 'any', 'emotion', 'akin', 'to', 'love', 'for',
'Irene', 'Adler', '.', 'All', 'emotions', ',', 'and', 'that',
'one', 'particularly', ',', 'were', 'abhorrent', 'to', 'his',
'cold', ',', 'precise', 'but', 'admirably', 'balanced', 'mind',
'.', 'He', 'was', ',', 'I', 'take', 'it', ',', 'the', 'most',
'perfect', 'reasoning', 'and', 'observing', 'machine', 'that',
'the', 'world', 'has', 'seen', ',', 'but', 'as', 'a', 'lover',
'he', 'would', 'have', 'placed', 'himself', 'in', 'a', 'false',
'position', '.', 'He', 'never', 'spoke', 'of', 'the', 'softer',
'passions', ',', 'save', 'with', 'a', 'gibe', 'and', 'a',
'sneer', '.', 'They', 'were', 'admirable', 'things', 'for',
'the', 'observer—excellent', 'for', 'drawing', 'the', 'veil',
'from', 'men', ''', 's', 'motives', 'and', 'actions', '.',
'But', 'for', 'the', 'trained', 'reasoner', 'to', 'admit',
'such', 'intrusions', 'into', 'his', 'own', 'delicate', 'and',
'finely', 'adjusted', 'temperament', 'was', 'to', 'introduce',
'a', 'distracting', 'factor', 'which', 'might', 'throw', 'a',
'doubt', 'upon', 'all', 'his', 'mental', 'results', '.', 'Grit',
'in', 'a', 'sensitive', 'instrument', ',', 'or', 'a', 'crack',
'in', 'one', 'of', 'his', 'own', 'high-power', 'lenses',
',', 'would', 'not', 'be', 'more', 'disturbing', 'than', 'a',
'strong', 'emotion', 'in', 'a', 'nature', 'such', 'as', 'his',
'.', 'And', 'yet', 'there', 'was', 'but', 'one', 'woman',
'to', 'him', ',', 'and', 'that', 'woman', 'was', 'the', 'late',
'Irene', 'Adler', ',', 'of', 'dubious', 'and', 'questionable',
'memory', '.']
230
```

The output is a list, where each token is either a word or a punctuation mark. The NLTK tokenizer uses a set of rules to split the text into words. It splits but does not expand contractions, such as *don't* → *do n't* and *men's* → *men 's*, as in the preceding example. It treats punctuation and quotes as separate tokens, so the result includes words with no other marks.

There's more...

Sometimes, it is useful not to split some words and use them as one unit. One example of this is in *Chapter 3*, in the *Representing phrases – phrase2vec* recipe, where we store phrases and not just individual words. The NLTK package allows us to do that using its custom tokenizer, MWETokenizer:

1. Import the MWETokenizer class:

```
from nltk.tokenize import MWETokenizer
```

2. Initialize the tokenizer and indicate that the words `dim sum dinner` should not be split:

```
tokenizer = MWETokenizer([('dim', 'sum', 'dinner')])
```

3. Add more words that should be kept together:

```
tokenizer.add_mwe(('best', 'dim', 'sum'))
```

4. Use the tokenizer to split a sentence:

```
tokens = tokenizer.tokenize('Last night I went for dinner in an
Italian restaurant. The pasta was delicious.'.split())
print(tokens)
```

The result will contain the tokens split the same way as previously:

```
['Last', 'night', 'I', 'went', 'for', 'dinner', 'in', 'an',
'Italian', 'restaurant.', 'The', 'pasta', 'was', 'delicious.']
```

5. Split a different sentence:

```
tokens = tokenizer.tokenize('I went out to a dim sum dinner last
night. This restaurant has the best dim sum in town.'.split())
print(tokens)
```

In this case, the tokenizer will put the phrases together into one unit and insert underscores instead of spaces:

```
['I', 'went', 'out', 'to', 'a', 'dim_sum_dinner', 'last',
'night.', 'This', 'restaurant', 'has', 'the_best_dim_sum', 'in',
'town.']
```

We can also use spaCy to do the tokenization. Word tokenization is one task in a larger array of tasks that spaCy accomplishes while processing text.

There's still more

If you are doing further processing on the text, it makes sense to use spaCy. Here is how it works:

1. Import the `spacy` package:

```
import spacy
```

2. Execute this command only if you have not before:

```
!python -m spacy download en_core_web_sm
```

3. Initialize the spaCy engine using the English model:

```
nlp = spacy.load("en_core_web_sm")
```

4. Divide the text into sentences:

```
doc = nlp(sherlock_holmes_part_of_text)
words = [token.text for token in doc]
```

5. Print the result:

```
print(words)
print(len(words))
```

The output will be as follows:

```
['To', 'Sherlock', 'Holmes', 'she', 'is', 'always', '_', 'the',
'_', 'woman', '.', 'I', 'have', 'seldom', 'heard', 'him', '\n',
'mention', 'her', 'under', 'any', 'other', 'name', '.', 'In',
'his', 'eyes', 'she', 'eclipses', 'and', '\n', 'predominates',
'the', 'whole', 'of', 'her', 'sex', '.', 'It', 'was', 'not',
'that', 'he', 'felt', 'any', 'emotion', '\n', 'akin', 'to',
'love', 'for', 'Irene', 'Adler', '.', 'All', 'emotions', ',',
'and', 'that', 'one', 'particularly', ',', '\n', 'were',
'abhorrent', 'to', 'his', 'cold', ',', 'precise', 'but',
'admirably', 'balanced', 'mind', '.', 'He', '\n', 'was', ',',
'I', 'take', 'it', ',', 'the', 'most', 'perfect', 'reasoning',
'and', 'observing', 'machine', 'that', '\n', 'the', 'world',
'has', 'seen', ',', 'but', 'as', 'a', 'lover', 'he', 'would',
'have', 'placed', 'himself', 'in', 'a', '\n', 'false',
'position', '.', 'He', 'never', 'spoke', 'of', 'the', 'softer',
'passions', ',', 'save', 'with', 'a', 'gibe', '\n', 'and', 'a',
'sneer', '.', 'They', 'were', 'admirable', 'things', 'for',
'the', 'observer', '—', 'excellent', 'for', '\n', 'drawing',
'the', 'veil', 'from', 'men', "'s", 'motives', 'and', 'actions',
'.', 'But', 'for', 'the', 'trained', '\n', 'reasoner', 'to',
'admit', 'such', 'intrusions', 'into', 'his', 'own', 'delicate',
'and', 'finely', '\n', 'adjusted', 'temperament', 'was',
'to', 'introduce', 'a', 'distracting', 'factor', 'which',
'might', '\n', 'throw', 'a', 'doubt', 'upon', 'all', 'his',
'mental', 'results', '.', 'Grit', 'in', 'a', 'sensitive', '\n',
'instrument', ',', 'or', 'a', 'crack', 'in', 'one', 'of',
'his', 'own', 'high', '-', 'power', 'lenses', ',', 'would',
'not', '\n', 'be', 'more', 'disturbing', 'than', 'a', 'strong',
'emotion', 'in', 'a', 'nature', 'such', 'as', 'his', '.', 'And',
'\n', 'yet', 'there', 'was', 'but', 'one', 'woman', 'to', 'him',
',', 'and', 'that', 'woman', 'was', 'the', 'late', 'Irene',
'\n', 'Adler', ',', 'of', 'dubious', 'and', 'questionable',
'memory', '.']
251
```

You will notice that the length of the word list is longer when using spaCy than NLTK. One of the reasons is that spaCy keeps the newlines, and each newline is a separate token. The other difference is that spaCy splits words with a dash, such as *high-power*. You can find the exact difference between the two lists by running the following line:

```
print(set(words_spacy)-set(words_nltk))
```

This should result in the following output:

```
{'high', 'power', 'observer', '-', '_', '—', 'excellent', ''s', '\n'}
```

> **Important note**
> If you are doing other processing with spaCy, it makes sense to use it. Otherwise, NLTK word tokenization is sufficient.

See also

The NLTK package only has word tokenization for English.

spaCy has models for other languages: Chinese, Dutch, English, French, German, Greek, Italian, Japanese, Portuguese, Romanian, Spanish, and others. In order to use those models, you would have to download them separately. For example, for Spanish, use this command to download the model:

```
python -m spacy download es_core_news_sm
```

Then, put this line in the code to use it:

```
nlp = spacy.load("es_core_news_sm")
```

See the spaCy documentation to find out more: https://spacy.io/usage/models.

Part of speech tagging

In many cases, NLP processing depends on determining the parts of speech of the words in the text. For example, when we want to find out the named entities that appear in a text, we need to know the parts of speech of the words. In this recipe, we will again consider NLTK and spaCy algorithms.

Getting ready

For this part, we will be using the same text of the book *The Adventures of Sherlock Holmes*. You can find the whole text in the book's Github repository. For this recipe, we will need just the beginning of the book, which can be found in the file at https://github.com/PacktPublishing/Python-Natural-Language-Processing-Cookbook-Second-Edition/blob/main/data/sherlock_holmes_1.txt.

In order to do this task, you will need the NLTK and spaCy packages, described in the *Technical requirements* section.

We will also complete this task using the OpenAI API's GPT model to demonstrate that it can complete it as well as spaCy and NLTK. For this part to run, you will need the openai package, which is included in the Poetry environment. You will also need your own OpenAI API key.

How to do it...

In this recipe, we will use the spaCy package to label words with their parts of speech.

The process is as follows:

1. Import the `util` file and the language `util` file. The language `util` file contains an import of spaCy and NLTK, as well as an initialization of the small spaCy model into the `small_model` object. These files also include functions to read in text from a file and tokenization functions using spaCy and NLTK:

   ```
   %run -i "../util/file_utils.ipynb"
   %run -i "../util/lang_utils.ipynb"
   ```

2. We will define the function that will output parts of speech for every word. In this function, we first process the input text using the spaCy model that results in a `Document` object. The resulting `Document` object contains an iterator with `Token` objects, and each `Token` object has information about parts of speech.

   We use this information to create the two lists, one with words and the other one with their respective parts of speech.

   Finally, we zip the two lists to pair the words with the parts of speech and return the resulting list of tuples. We do this in order to easily print the whole list with their corresponding parts of speech. When you use part of speech tagging in your code, you can just iterate through the list of tokens:

   ```
   def pos_tag_spacy(text, model):
       doc = model(text)
       words = [token.text for token in doc]
       pos = [token.pos_ for token in doc]
       return list(zip(words, pos))
   ```

3. Read in the text:

   ```
   text = read_text_file("../data/sherlock_holmes_1.txt")
   ```

4. Run the preceding function using the text and the model as input:

   ```
   words_with_pos = pos_tag_spacy(text, small_model)
   ```

5. Print the output:

   ```
   print(words_with_pos)
   ```

   Part of the result is shown in the following; for the complete output, please see the Jupyter notebook (https://github.com/PacktPublishing/Python-Natural-Language-

Processing-Cookbook-Second-Edition/blob/main/Chapter01/part_of_
speech_tagging_1.3.ipynb):

```
[('To', 'ADP'),
 ('Sherlock', 'PROPN'),
 ('Holmes', 'PROPN'),
 ('she', 'PRON'),
 ('is', 'AUX'),
 ('always', 'ADV'),
 ('_', 'PUNCT'),
 ('the', 'DET'),
 ('_', 'PROPN'),
 ('woman', 'NOUN'),
 ('.', 'PUNCT'),
 ('I', 'PRON'),
 ('have', 'AUX'),
 ('seldom', 'ADV'),
 ('heard', 'VERB'),
 ('him', 'PRON'),
 ('\n', 'SPACE'),
 ('mention', 'VERB'),
 ('her', 'PRON'),
 ('under', 'ADP'),
 ('any', 'DET'),
 ('other', 'ADJ'),
 ('name', 'NOUN'),
 ('.', 'PUNCT'),…
```

The resulting list contains tuples of words and parts of speech. The list of part of speech tags is available here: https://universaldependencies.org/u/pos/.

There's more

We can compare spaCy's performance to NLTK in this task. Here are the steps for getting the parts of speech with NLTK:

1. The imports have been taken care of in the language util file that we imported, so the first thing we do is create a function that outputs parts of speech for the words that are input. In it, we utilize the word_tokenize_nltk function that is also imported from the language util notebook:

```
def pos_tag_nltk(text):
    words = word_tokenize_nltk(text)
    words_with_pos = nltk.pos_tag(words)
    return words_with_pos
```

2. Next, we apply the function to the text that we read in previously:

```
words_with_pos = pos_tag_nltk(text)
```

3. Print out the result:

```
print(words_with_pos)
```

Part of the output is shown in the following. For the complete output, please see the Jupyter notebook:

```
[('To', 'TO'),
 ('Sherlock', 'NNP'),
 ('Holmes', 'NNP'),
 ('she', 'PRP'),
 ('is', 'VBZ'),
 ('always', 'RB'),
 ('_the_', 'JJ'),
 ('woman', 'NN'),
 ('.', '.'),
 ('I', 'PRP'),
 ('have', 'VBP'),
 ('seldom', 'VBN'),
 ('heard', 'RB'),
 ('him', 'PRP'),
 ('mention', 'VB'),
 ('her', 'PRP'),
 ('under', 'IN'),
 ('any', 'DT'),
 ('other', 'JJ'),
 ('name', 'NN'),
 ('.', '.'),…
```

The list of part of speech tags that NLTK uses is different from what SpaCy uses, and can be accessed by running the following commands:

```
python
>>> import nltk
>>> nltk.download('tagsets')
>>> nltk.help.upenn_tagset()
```

Comparing the performance, we see that spaCy takes 0.02 seconds, while NLTK takes 0.01 seconds (your numbers might be different), so their performance is similar, with NLTK being a little better. However, the part of speech information is already available in the spaCy objects after the initial processing has been done, so if you are doing any further processing, spaCy is a better choice.

> **Important note**
> spaCy does all of its processing at once, and the results are stored in the `Doc` object. The part
> of speech information is available by iterating through `Token` objects.

There's more

We can use the OpenAI API with the GPT-3.5 and GPT-4 models to perform various tasks, including
many NLP ones. Here, we show how to use the OpenAI API to get NLTK-style parts of speech for
input text. You can also specify in the prompt the output format and the style of the part of speech
tags. For this code to run correctly, you will need your own OpenAI API key:

1. Import `openai` and create the OpenAI client using your API key. The `OPEN_AI_KEY` constant
    variable is set in the `../util/file_utils.ipynb` file:

    ```
    from openai import OpenAI
    client = OpenAI(api_key=OPEN_AI_KEY)
    ```

2. Set the prompt:

    ```
    prompt="""Decide what the part of speech tags are for a
    sentence.
    Preserve original capitalization.
    Return the list in the format of a python tuple: (word, part of
    speech).
    Sentence: In his eyes she eclipses and predominates the whole of
    her sex."""
    ```

3. Send the request to the OpenAI API. Some of the important parameters that we send to the API
    are the model we want to use, the temperature, which affects how much the response from the
    model will vary, and the maximum amount of tokens the model should return as a completion:

    ```
    response = client.chat.completions.create(
        model="gpt-3.5-turbo",
        temperature=0,
        max_tokens=256,
        top_p=1.0,
        frequency_penalty=0,
        presence_penalty=0,
        messages=[
            {"role": "system",
             "content": "You are a helpful assistant."},
            {"role": "user", "content": prompt}
        ],
    )
    ```

4. Print the response:

```
print(response)
```

The output will look like this:

```
ChatCompletion(id='chatcmpl-9hCq34UAzMiNiqNGopt2U8ZmZM5po',
choices=[Choice(finish_reason='stop', index=0, logprobs=None,
message=ChatCompletionMessage(content='Here are the part of
speech tags for the sentence "In his eyes she eclipses and
predominates the whole of her sex" in the format of a Python
tuple:\n\n[(\'In\', \'IN\'), (\'his\', \'PRP$\'), (\'eyes\',
\'NNS\'), (\'she\', \'PRP\'), (\'eclipses\', \'VBZ\'), (\'and\',
\'CC\'), (\'predominates\', \'VBZ\'), (\'the\', \'DT\'),
(\'whole\', \'JJ\'), (\'of\', \'IN\'), (\'her\', \'PRP$\'),
(\'sex\', \'NN\')]', role='assistant', function_call=None, tool_
calls=None))], created=1720084483, model='gpt-3.5-turbo-0125',
object='chat.completion', service_tier=None, system_
fingerprint=None, usage=CompletionUsage(completion_tokens=120,
prompt_tokens=74, total_tokens=194))
```

5. To see just the GPT output, do the following:

```
print(response.choices[0].message.content)
```

The output will be as follows:

```
Here are the part of speech tags for the sentence "In his eyes
she eclipses and predominates the whole of her sex" in the
format of a Python tuple:

[('In', 'IN'), ('his', 'PRP$'), ('eyes', 'NNS'), ('she', 'PRP'),
('eclipses', 'VBZ'), ('and', 'CC'), ('predominates', 'VBZ'),
('the', 'DT'), ('whole', 'JJ'), ('of', 'IN'), ('her', 'PRP$'),
('sex', 'NN')]
```

6. We can use the `literal_eval` function to transform the response into a tuple. We request that the GPT model return only the answer without additional explanations so that there is no free text inside the answer and we can process it automatically. We do this in order to be able to compare the output of the OpenAI API to the NLTK output:

```
from ast import literal_eval


def pos_tag_gpt(text, client):
    prompt = f"""Decide what the part of speech tags are for a
sentence.
    Preserve original capitalization.
    Return the list in the format of a python tuple: (word, part
of speech).
    Do not include any other explanations.
    Sentence: {text}."""


    response = client.chat.completions.create(
```

```
        model="gpt-3.5-turbo",
        temperature=0,
        max_tokens=256,
        top_p=1.0,
        frequency_penalty=0,
        presence_penalty=0,
        messages=[
            {"role": "system",
             "content": "You are a helpful assistant."},
            {"role": "user", "content": prompt}
        ],
    )
    result = response.choices[0].message.content
    result = result.replace("\n", "")
    result = list(literal_eval(result))
    return result
```

7. Now, let's time the GPT function so we can compare its performance to the other methods we used previously:

```
start = time.time()
first_sentence = "In his eyes she eclipses and predominates the
whole of her sex."
words_with_pos = pos_tag_gpt(first_sentence, OPEN_AI_KEY)
print(words_with_pos)
print(f"GPT: {time.time() - start} s")
```

The result will look like this:

```
[('In', 'IN'), ('his', 'PRP$'), ('eyes', 'NNS'), ('she', 'PRP'),
('eclipses', 'VBZ'), ('and', 'CC'), ('predominates', 'VBZ'),
('the', 'DT'), ('whole', 'NN'), ('of', 'IN'), ('her', 'PRP$'),
('sex', 'NN'), ('.', '.')]
GPT: 2.4942469596862793 s
```

8. The output of GPT is very similar to NLTK, but slightly different:

```
words_with_pos_nltk = pos_tag_nltk(first_sentence)
print(words_with_pos == words_with_pos_nltk)
```

This outputs the following:

```
False
```

The difference between GPT and NLTK is that GPT tags the word whole as an adjective and NLTK tags it as a noun. In this context, NLTK is correct.

We see that the LLM outputs very similar results but is about 400 times slower than NLTK.

See also

If you would like to tag a text in another language, you can do so by using spaCy's models for other languages. For example, we can load the Spanish spaCy model to run it on Spanish text:

```
nlp = spacy.load("es_core_news_sm")
```

In the case that spaCy doesn't have a model for the language you are working with, you can train your own model with spaCy. See https://spacy.io/usage/training#tagger-parser.

Combining similar words – lemmatization

We can find the canonical form of the word using **lemmatization**. For example, the lemma of the word *cats* is *cat*, and the lemma for the word *ran* is *run*. This is useful when we are trying to match some word and don't want to list out all the possible forms. Instead, we can just use its lemma.

Getting ready

We will be using the spaCy package for this recipe.

How to do it...

When the spaCy model processes a piece of text, the resulting Document object contains an iterator over the Token objects within it, as we saw in the *Part of speech tagging* recipe. These Token objects contain the lemma information for each word in the text.

Here are the steps for getting the lemmas:

1. Import the file and language utils files. This will import spaCy and initialize the small_ model object:

    ```
    %run -i "../util/file_utils.ipynb"
    %run -i "../util/lang_utils.ipynb"
    ```

2. Create a list of words we want to lemmatize:

    ```
    words = ["leaf", "leaves", "booking", "writing", "completed",
    "stemming"]
    ```

3. Create a Document object for each of the words:

    ```
    docs = [small_model(word) for word in words]
    ```

4. Print the words and their lemmas for each of the words in the list:

    ```
    for doc in docs:
        for token in doc:
            print(token, token.lemma_)
    ```

The result will be as follows:

```
leaf leaf
leaves leave
booking book
writing write
completed complete
stemming stem
```

The result shows correct lemmatization for all words. However, some words are ambiguous. For example, the word *leaves* could either be a verb, in which case the lemma is correct, or a noun, in which case this is the wrong lemma. If we give the spaCy continuous text instead of individual words, it is likely to correctly disambiguate the words.

5. Now, apply lemmatization to the longer text. Here, we read in a small portion of the *Sherlock Holmes* text and lemmatize its every word:

```
Text = read_text_file(../data/sherlock_holmes_1.txt")
doc = small_model(text)
for token in doc:
    print(token, token.lemma_)
```

The partial result will be as follows:

```
To to
Sherlock Sherlock
Holmes Holmes
she she
is be
always always

— —
the the

— —
woman woman
. ….
```

There's more...

We can use the spaCy lemmatizer object to find out whether a word is in its base form or not. We might do this while manipulating the grammar of the sentence, for example, in the task of turning a passive sentence into an active one. We can get to the lemmatizer object by manipulating the spaCy pipeline, which includes various tools that are applied to the text. See `https://spacy.io/usage/processing-pipelines/` for more information. Here are the steps:

1. The pipeline components are located in a list of tuples, (`component name`, `component`). To get the lemmatizer component, we need to loop through this list:

```
lemmatizer = None
for name, proc in small_model.pipeline:
    if name == "lemmatizer":
        lemmatizer = proc
```

2. Now, we can apply the `is_base_form` function call to every word in the Sherlock Holmes text:

```
for token in doc:
    print(f"{token} is in its base form:
        {lemmatizer.is_base_form(token)}")
```

The partial result will be as follows:

```
To is in its base form: False
Sherlock is in its base form: False
Holmes is in its base form: False
she is in its base form: False
is is in its base form: False
always is in its base form: False
_ is in its base form: False
the is in its base form: False
_ is in its base form: False
woman is in its base form: True
. is in its base form: False…
```

Removing stopwords

When we work with words, especially if we are considering the words' semantics, we sometimes need to exclude some very frequent words that do not bring any substantial meaning into the sentence (words such as *but*, *can*, *we*, etc.). For example, if we want to get a rough sense of the topic of a text, we could count its most frequent words. However, in any text, the most frequent words will be stopwords, so we want to remove them before processing. This recipe shows how to do that. The stopwords list we are using in this recipe comes from the NLTK package and might not include all the words you need. You will need to modify the list accordingly.

Getting ready

We will remove stopwords using spaCy and NLTK; these packages are part of the Poetry environment that we installed earlier.

We will be using the *Sherlock Holmes* text referred to earlier. For this recipe, we will need just the beginning of the book, which can be found in the file at `https://github.com/PacktPublishing/Python-Natural-Language-Processing-Cookbook-Second-Edition/blob/main/data/sherlock_holmes_1.txt`.

In *step 1*, we run the utilities notebooks. In *step 2*, we import the `nltk` package and its stopwords list. In *step 3*, we download the stopwords data, if necessary. In *step 4*, we print out the stopwords list. In *step 5*, we read in a small portion of the *Sherlock Holmes* book. In *step 6*, we tokenize the text and print its length, which is 230. In *step 7*, we remove the stopwords from the original words list by using a list comprehension. Then, we print the length of the result and see that the list length has been reduced to 105. You will notice that in the list comprehension, we check whether the *lowercase* version of the word is in the stopwords list since all the stopwords are lowercase.

How to do it...

In the recipe, we will read in the text file, tokenize the text, and remove the stopwords from the list:

1. Run the file and language utilities notebooks:

    ```
    %run -i "../util/file_utils.ipynb"
    %run -i "../util/lang_utils.ipynb"
    ```

2. Import the NLTK stopwords list:

    ```
    from nltk.corpus import stopwords
    ```

3. The first time you run the notebook, download the `stopwords` data. You don't need to download the stopwords again the next time you run the code:

    ```
    nltk.download('stopwords')
    ```

> **Note**
>
> Here is a list of languages that NLTK supports for stopwords: Arabic, Azerbaijani, Danish, Dutch, English, Finnish, French, German, Greek, Hungarian, Italian, Kazakh, Nepali, Norwegian, Portuguese, Romanian, Russian, Spanish, Swedish, and Turkish.

4. You can see the stopwords that come with NLTK by printing the list:

    ```
    print(stopwords.words('english'))
    ```

    The result will be as follows:

    ```
    ['i', 'me', 'my', 'myself', 'we', 'our', 'ours', 'ourselves',
    'you', "you're", "you've", "you'll", "you'd", 'your', 'yours',
    'yourself', 'yourselves', 'he', 'him', 'his', 'himself',
    'she', "she's", 'her', 'hers', 'herself', 'it', "it's", 'its',
    'itself', 'they', 'them', 'their', 'theirs', 'themselves',
    ```

```
'what', 'which', 'who', 'whom', 'this', 'that', "that'll",
'these', 'those', 'am', 'is', 'are', 'was', 'were', 'be',
'been', 'being', 'have', 'has', 'had', 'having', 'do', 'does',
'did', 'doing', 'a', 'an', 'the', 'and', 'but', 'if', 'or',
'because', 'as', 'until', 'while', 'of', 'at', 'by', 'for',
'with', 'about', 'against', 'between', 'into', 'through',
'during', 'before', 'after', 'above', 'below', 'to', 'from',
'up', 'down', 'in', 'out', 'on', 'off', 'over', 'under',
'again', 'further', 'then', 'once', 'here', 'there', 'when',
'where', 'why', 'how', 'all', 'any', 'both', 'each', 'few',
'more', 'most', 'other', 'some', 'such', 'no', 'nor', 'not',
'only', 'own', 'same', 'so', 'than', 'too', 'very', 's', 't',
'can', 'will', 'just', 'don', "don't", 'should', "should've",
'now', 'd', 'll', 'm', 'o', 're', 've', 'y', 'ain', 'aren',
"aren't", 'couldn', "couldn't", 'didn', "didn't", 'doesn',
"doesn't", 'hadn', "hadn't", 'hasn', "hasn't", 'haven',
"haven't", 'isn', "isn't", 'ma', 'mightn', "mightn't", 'mustn',
"mustn't", 'needn', "needn't", 'shan', "shan't", 'shouldn',
"shouldn't", 'wasn', "wasn't", 'weren', "weren't", 'won',
"won't", 'wouldn', "wouldn't"]
```

5. Read in the text file:

    ```
    text = read_text_file("../data/sherlock_holmes_1.txt")
    ```

6. Tokenize the text and print the length of the resulting list:

    ```
    words = word_tokenize_nltk(text)
    print(len(words))
    ```

    The result will be as follows:

    ```
    230
    ```

7. Remove the stopwords from the list using a list comprehension and print the length of the result. You will notice that in the list comprehension, we check whether the *lowercase* version of the word is in the stopwords list since all the stopwords are lowercase.

    ```
    words = [word for word in words if word not in stopwords.
    words("english")]
    print(len(words))
    ```

    The result will be as follows:

    ```
    105
    ```

The code then filters the stopwords from the text and leaves the words from the text only if they do not also appear in the stopwords list. As we see from the lengths of the two lists, one unfiltered and the other without the stopwords, we remove more than half the words.

> **Important note**
> You might find that some of the words in the stopwords list provided are not necessary or are missing. You will need to modify the list accordingly. The NLTK stopwords list is a Python list and you can add and remove elements using the standard Python list functions.

There's more...

We can also remove stopwords using spaCy. Here is how to do it:

1. Assign the stopwords to a variable for convenience:

```
stopwords = small_model.Defaults.stop_words
```

2. Tokenize the text and print its length:

```
words = word_tokenize_nltk(text)
print(len(words))
```

It will give us the following result:

```
230
```

3. Remove the stopwords from the list using a list comprehension and print the resulting length:

```
words = [word for word in words if word.lower() not in
stopwords]
print(len(words))
```

The result will be very similar to NLTK :

```
106
```

4. The stopwords from spaCy are stored in a set and we can add more words to it:

```
print(len(stopwords))
stopwords.add("new")
print(len(stopwords))
```

The result will be as follows:

```
327
328
```

Similarly, we can remove words if necessary:

```
print(len(stopwords))
stopwords.remove("new")
print(len(stopwords))
```

The result will be as follows:

```
328
327
```

We can also compile a stopwords list using the text we are working with and calculate the frequencies of the words in it. This provides you with an automatic way of removing stopwords, without the need for manual review.

There's still more

In this section, we will show you two ways of doing so. You will need to use the file at https://github.com/PacktPublishing/Python-Natural-Language-Processing-Cookbook-Second-Edition/blob/main/data/sherlock_holmes.txt. The FreqDist object in the NLTK package counts the number of occurrences of each word that we later use to find the most frequent words and remove them as stopwords:

1. Import the NTLK FreqDist class:

    ```
    from nltk.probability import FreqDist
    ```

2. Define the function that will compile a list of stopwords:

    ```
    def compile_stopwords_list_frequency(text, cut_off=0.02):
        words = word_tokenize_nltk(text)
        freq_dist = FreqDist(word.lower() for word in words)
        words_with_frequencies = [
            (word, freq_dist[word]) for word in freq_dist.keys()]
        sorted_words = sorted(words_with_frequencies,
            key=lambda tup: tup[1])
        stopwords = []
        if (type(cut_off) is int):
            # First option: use a frequency cutoff
            stopwords = [tuple[0] for tuple in sorted_words
                if tuple[1] > cut_off]

        elif (type(cut_off) is float):
            # Second option: use a percentage of the words
            length_cutoff = int(cut_off*len(sorted_words))
            stopwords = [tuple[0] for tuple in
                sorted_words[-length_cutoff:]]

        else:
            raise TypeError("The cut off needs to be either a float
    (percentage) or an int (frequency cut off)")
        return stopwords
    ```

3. Define the stopwords list using the default settings, and print out the result and its length:

```
text = read_text_file("../data/sherlock_holmes.txt")
stopwords = compile_stopwords_list_frequency(text)
print(stopwords)
print(len(stopwords))
```

The result will be as follows:

```
['make', 'myself', 'night', 'until', 'street', 'few', 'why',
'thought', 'take', 'friend', 'lady', 'side', 'small', 'still',
'these', 'find', 'st.', 'every', 'watson', 'too', 'round',
'young', 'father', 'left', 'day', 'yet', 'first', 'once',
'took', 'its', 'eyes', 'long', 'miss', 'through', 'asked',
'most', 'saw', 'oh', 'morning', 'right', 'last', 'like', 'say',
'tell', 't', 'sherlock', 'their', 'go', 'own', 'after', 'away',
'never', 'good', 'nothing', 'case', 'however', 'quite', 'found',
'made', 'house', 'such', 'heard', 'way', 'yes', 'hand', 'much',
'matter', 'where', 'might', 'just', 'room', 'any', 'face',
'here', 'back', 'door', 'how', 'them', 'two', 'other', 'came',
'time', 'did', 'than', 'come', 'before', 'must', 'only', 'know',
'about', 'shall', 'think', 'more', 'over', 'us', 'well', 'am',
'or', 'may', 'they', ';', 'our', 'should', 'now', 'see', 'down',
'can', 'some', 'if', 'will', 'mr.', 'little', 'who', 'into',
'do', 'has', 'could', 'up', 'man', 'out', 'when', 'would', 'an',
'are', 'by', '!', 'were', 's', 'then', 'one', 'all', 'on', 'no',
'what', 'been', 'your', 'very', 'him', 'her', 'she', 'so', "'",
'holmes', 'upon', 'this', 'said', 'from', 'there', 'we', 'me',
'be', 'but', 'not', 'for', '?', 'at', 'which', 'with', 'had',
'as', 'have', 'my', "'", 'is', 'his', 'was', 'you', 'he', 'it',
'that', 'in', '"', 'a', 'of', 'to', '"', 'and', 'i', '.', 'the',
',']
181
```

4. Now, use the function with the frequency cut-off of 5% (use the top 5% of the most frequent words as stopwords):

```
text = read_text_file("../data/sherlock_holmes.txt")
stopwords = compile_stopwords_list_frequency(text, cut_off=0.05)
print(len(stopwords))
```

The result will be as follows:

```
452
```

5. Now, use the absolute frequency cut-off of 100 (take the words that have a frequency greater than 100):

```
stopwords = compile_stopwords_list_frequency(text, cut_off=100)
print(stopwords)
print(len(stopwords))
```

And the result is the following:

```
['away', 'never', 'good', 'nothing', 'case', 'however', 'quite',
'found', 'made', 'house', 'such', 'heard', 'way', 'yes', 'hand',
'much', 'matter', 'where', 'might', 'just', 'room', 'any',
'face', 'here', 'back', 'door', 'how', 'them', 'two', 'other',
'came', 'time', 'did', 'than', 'come', 'before', 'must', 'only',
'know', 'about', 'shall', 'think', 'more', 'over', 'us', 'well',
'am', 'or', 'may', 'they', ';', 'our', 'should', 'now', 'see',
'down', 'can', 'some', 'if', 'will', 'mr.', 'little', 'who',
'into', 'do', 'has', 'could', 'up', 'man', 'out', 'when',
'would', 'an', 'are', 'by', '!', 'were', 's', 'then', 'one',
'all', 'on', 'no', 'what', 'been', 'your', 'very', 'him', 'her',
'she', 'so', "'", 'holmes', 'upon', 'this', 'said', 'from',
'there', 'we', 'me', 'be', 'but', 'not', 'for', '?', 'at',
'which', 'with', 'had', 'as', 'have', 'my', "'", 'is', 'his',
'was', 'you', 'he', 'it', 'that', 'in', '"', 'a', 'of', 'to',
'"', 'and', 'i', '.', 'the', ',']
131
```

The function that creates the stopwords list takes in the text and the cut_off parameter. It could be a float representing the percentage of frequency-ranked words that will be in the stopwords list. Alternatively, it could be an integer that represents the absolute threshold frequency, with words above it considered stopwords. In the function, we first tokenize the words from the book, then create a FreqDist object, and then create a list of tuples (word, word's frequency) using the frequency distribution. We sort the list using the word frequency. We then check the cut_off parameter's type and raise an error if it is not a float or an integer. If it is an integer, we return all words whose frequency is higher than the parameter as stopwords. If it is a float, we calculate the number of words to be returned using the parameter as the percentage.

2
Playing with Grammar

Grammar is one of the main building blocks of language. Each human language, and programming language for that matter, has a set of rules that every person speaking it must follow, otherwise risking not being understood. These grammatical rules can be uncovered using NLP and are useful for extracting data from sentences. For example, using information about the grammatical structure of text, we can parse out subjects, objects, and relations between different entities.

In this chapter, you will learn how to use different packages to reveal the grammatical structure of words and sentences, as well as extract certain parts of sentences. These are the topics covered in this chapter:

- Counting nouns – plural and singular nouns
- Getting the dependency parse
- Extracting noun chunks
- Extracting the subjects and objects of the sentence
- Finding patterns in text using grammatical information

Technical requirements

Please follow the installation requirements given in *Chapter 1* to run the notebooks in this chapter.

Counting nouns – plural and singular nouns

In this recipe, we will do two things: determine whether a noun is plural or singular and turn plural nouns into singular, and vice versa.

You might need these two things for a variety of tasks. For example, you might want to count the word statistics, and for that, you most likely need to count the singular and plural nouns together. In order to count the plural nouns together with singular ones, you need a way to recognize that a word is plural or singular.

Getting ready

To determine whether a noun is singular or plural, we will use spaCy via two different methods: by looking at the difference between the lemma and the actual word and by looking at the morph attribute. To inflect these nouns, or turn singular nouns into plural or vice versa we will use the textblob package. We will also see how to determine the noun's number using GPT-3 through the OpenAI API. The code for this section is located at https://github.com/PacktPublishing/Python-Natural-Language-Processing-Cookbook-Second-Edition/tree/main/Chapter02.

How to do it...

We will first use spaCy's lemma information to infer whether a noun is singular or plural. Then, we will use the morph attribute of Token objects. We will then create a function that uses one of those methods. Finally, we will use GPT-3.5 to find out the number of nouns:

1. Run the code in the file and language utility notebooks. If you run into an error saying that the small or large models do not exist, you need to open the lang_utils.ipynb file, uncomment, and run the statement that downloads the model:

```
%run -i "../util/file_utils.ipynb"
%run -i "../util/lang_utils.ipynb"
```

2. Initialize the text variable and process it using the spaCy small model to get the resulting Doc object:

```
text = "I have five birds"
doc = small_model(text)
```

3. In this step, we loop through the Doc object. For each token in the object, we check whether it's a noun and whether the lemma is the same as the word itself. Since the lemma is the basic form of the word, if the lemma is different from the word, that token is plural:

```
for token in doc:
    if (token.pos_ == "NOUN" and token.lemma_ != token.text):
        print(token.text, "plural")
```

The result should be as follows:

```
birds plural
```

4. Now, we will check the number of a noun using a different method: the morph features of a Token object. The morph features are the morphological features of a word, such as number, case, and so on. Since we know that token 3 is a noun, we directly access the morph features and get the Number to get the same result as previously:

```
doc = small_model("I have five birds.")
print(doc[3].morph.get("Number"))
```

Here is the result:

```
['Plur']
```

5. In this step, we prepare to define a function that returns a tuple, (noun, number). In order to better encode the noun number, we use an Enum class that assigns numbers to different values. We assign 1 to singular and 2 to plural. Once we create the class, we can directly refer to the noun number variables as Noun_number.SINGULAR and Noun_number.PLURAL:

```
class Noun_number(Enum):
    SINGULAR = 1
    PLURAL = 2
```

6. In this step, we define the function. It takes as input the text, the spaCy model, and the method of determining the noun number. The two methods are lemma and morph, the same two methods we used in *steps 3* and *4*, respectively. The function outputs a list of tuples, each of the format (<noun text>, <noun number>), where the noun number is expressed using the Noun_number class defined in *step 5*:

```
def get_nouns_number(text, model, method="lemma"):
    nouns = []
    doc = model(text)
    for token in doc:
        if (token.pos_ == "NOUN"):
            if method == "lemma":
                if token.lemma_ != token.text:
                    nouns.append((token.text,
                        Noun_number.PLURAL))
                else:
                    nouns.append((token.text,
                        Noun_number.SINGULAR))
            elif method == "morph":
                if token.morph.get("Number") == "Sing":
                    nouns.append((token.text,
                        Noun_number.PLURAL))
                else:
                    nouns.append((token.text,
                        Noun_number.SINGULAR))
    return nouns
```

7. We can use the preceding function and see its performance with different spaCy models. In this step, we use the small spaCy model with the function we just defined. Using both methods, we see that the spaCy model gets the number of the irregular noun geese incorrectly:

```
text = "Three geese crossed the road"
nouns = get_nouns_number(text, small_model, "morph")
```

```
print(nouns)
nouns = get_nouns_number(text, small_model)
print(nouns)
```

The result should be as follows:

```
[('geese', <Noun_number.SINGULAR: 1>), ('road', <Noun_number.
SINGULAR: 1>)]
[('geese', <Noun_number.SINGULAR: 1>), ('road', <Noun_number.
SINGULAR: 1>)]
```

8. Now, let's do the same using the large model. If you have not yet downloaded the large model, do so by running the first line. Otherwise, you can comment it out. Here, we see that although the morph method still incorrectly assigns singular to geese, the lemma method provides the correct answer:

```
!python -m spacy download en_core_web_lg
large_model = spacy.load("en_core_web_lg")
nouns = get_nouns_number(text, large_model, "morph")
print(nouns)
nouns = get_nouns_number(text, large_model)
print(nouns)
```

The result should be as follows:

```
[('geese', <Noun_number.SINGULAR: 1>), ('road', <Noun_number.
SINGULAR: 1>)]
[('geese', <Noun_number.PLURAL: 2>), ('road', <Noun_number.
SINGULAR: 1>)]
```

9. Let's now use GPT-3.5 to get the noun number. In the results, we see that GPT-3.5 gives us an identical result and correctly identifies both the number for geese and the number for road:

```
from openai import OpenAI
client = OpenAI(api_key=OPEN_AI_KEY)
prompt="""Decide whether each noun in the following text is
singular or plural.
Return the list in the format of a python tuple: (word, number).
Do not provide any additional explanations.
Sentence: Three geese crossed the road."""
response = client.chat.completions.create(
    model="gpt-3.5-turbo",
    temperature=0,
    max_tokens=256,
    top_p=1.0,
    frequency_penalty=0,
    presence_penalty=0,
    messages=[
```

```
            {"role": "system", "content": "You are a helpful
                assistant."},
            {"role": "user", "content": prompt}
        ],
    )
    print(response.choices[0].message.content)
```

The result should be as follows:

```
('geese', 'plural')
('road', 'singular')
```

There's more…

We can also change the nouns from plural to singular, and vice versa. We will use the `textblob` package for that. The package should be installed automatically via the Poetry environment:

1. Import the `TextBlob` class from the package:

    ```
    from textblob import TextBlob
    ```

2. Initialize a list of text variables and process them using the `TextBlob` class via a list comprehension:

    ```
    texts = ["book", "goose", "pen", "point", "deer"]
    blob_objs = [TextBlob(text) for text in texts]
    ```

3. Use the `pluralize` function of the object to get the plural. This function returns a list and we access its first element. Print the result:

    ```
    plurals = [blob_obj.words.pluralize()[0]
        for blob_obj in blob_objs]
    print(plurals)
    ```

    The result should be as follows:

    ```
    ['books', 'geese', 'pens', 'points', 'deer']
    ```

4. Now, we will do the reverse. We use the preceding `plurals` list to turn the plural nouns into `TextBlob` objects:

    ```
    blob_objs = [TextBlob(text) for text in plurals]
    ```

5. Turn the nouns into singular using the `singularize` function and print:

    ```
    singulars = [blob_obj.words.singularize()[0]
        for blob_obj in blob_objs]
    print(singulars)
    ```

    The result should be the same as the list we started with in *step 2*:

    ```
    ['book', 'goose', 'pen', 'point', 'deer']
    ```

Getting the dependency parse

A dependency parse is a tool that shows dependencies in a sentence. For example, in the sentence *The cat wore a hat*, the root of the sentence is the verb, *wore*, and both the subject, *the cat*, and the object, *a hat*, are dependents. The dependency parse can be very useful in many NLP tasks since it shows the grammatical structure of the sentence, with the subject, the main verb, the object, and so on. It can then be used in downstream processing.

The `spaCy` NLP engine does the dependency parse as part of its overall analysis. The dependency parse tags explain the role of each word in the sentence. ROOT is the main word that all other words depend on, usually the verb.

Getting ready

We will use `spaCy` to create the dependency parse. The required packages are part of the Poetry environment.

How to do it...

We will take a few sentences from the `sherlock_holmes1.txt` file to illustrate the dependency parse. The steps are as follows:

1. Run the file and language utility notebooks:

   ```
   %run -i "../util/file_utils.ipynb"
   %run -i "../util/lang_utils.ipynb"
   ```

2. Define the sentence we will be parsing:

   ```
   sentence = 'I have seldom heard him mention her under any other
   name.'
   ```

3. Define a function that will print the word, its grammatical function embedded in the `dep_` attribute, and the explanation of that attribute. The `dep_` attribute of the `Token` object shows the grammatical function of the word in the sentence:

   ```
   def print_dependencies(sentence, model):
       doc = model(sentence)
       for token in doc:
           print(token.text, "\t", token.dep_, "\t",
               spacy.explain(token.dep_))
   ```

4. Now, let's use this function on the first sentence in our list. We can see that the verb `heard` is the ROOT word of the sentence, with all other words depending on it:

   ```
   print_dependencies(sentence, small_model)
   ```

The result should be as follows:

```
I     nsubj      nominal subject
have     aux     auxiliary
seldom     advmod      adverbial modifier
heard     ROOT     root
him     nsubj      nominal subject
mention     ccomp      clausal complement
her     dobj      direct object
under     prep      prepositional modifier
any     det     determiner
other     amod      adjectival modifier
name     pobj      object of preposition
.     punct     punctuation
```

5. To explore the dependency parse structure, we can use the attributes of the Token class. Using the ancestors and children attributes, we can get the tokens that this token depends on and the tokens that depend on it, respectively. The function to print the ancestors is as follows:

```
def print_ancestors(sentence, model):
    doc = model(sentence)
    for token in doc:
        print(token.text, [t.text for t in token.ancestors])
```

6. Now, let's use this function on the first sentence in our list:

```
print_ancestors(sentence, small_model)
```

The output will be as follows. In the result, we see that heard has no ancestors since it is the main word in the sentence. All other words depend on it, and in fact, contain heard in their ancestor lists.

The dependency chain can be seen by following the ancestor links for each word. For example, if we look at the word name, we see that its ancestors are under, mention, and heard. The immediate parent of name is under, the parent of under is mention, and the parent of mention is heard. A dependency chain will always lead to the root, or the main word, of the sentence:

```
I ['heard']
have ['heard']
seldom ['heard']
heard []
him ['mention', 'heard']
mention ['heard']
her ['mention', 'heard']
under ['mention', 'heard']
any ['name', 'under', 'mention', 'heard']
```

```
other ['name', 'under', 'mention', 'heard']
name ['under', 'mention', 'heard']
. ['heard']
```

7. To see all the children, use the following function. This function prints out each word and the words that depend on it, its **children**:

```
def print_children(sentence, model):
    doc = model(sentence)
    for token in doc:
        print(token.text, [t.text for t in token.children])
```

8. Now, let's use this function on the first sentence in our list:

```
print_children(sentence, small_model)
```

The result should be as follows. Now, the word heard has a list of words that depend on it since it is the main word in the sentence:

```
I []
have []
seldom []
heard ['I', 'have', 'seldom', 'mention', '.']
him []
mention ['him', 'her', 'under']
her []
under ['name']
any []
other []
name ['any', 'other']
. []
```

9. We can also see left and right children in separate lists. In the following function, we print the children as two separate lists, left and right. This can be useful when doing grammatical transformations in the sentence:

```
def print_lefts_and_rights(sentence, model):
    doc = model(sentence)
    for token in doc:
        print(token.text,
                [t.text for t in token.lefts],
                [t.text for t in token.rights])
```

10. Let's use this function on the first sentence in our list:

```
print_lefts_and_rights(sentence, small_model)
```

The result should be as follows:

```
I [] []
have [] []
seldom [] []
heard ['I', 'have', 'seldom'] ['mention', '.']
him [] []
mention ['him'] ['her', 'under']
her [] []
under [] ['name']
any [] []
other [] []
name ['any', 'other'] []
. [] []
```

11. We can also see the subtree that the token is in by using this function:

```
def print_subtree(sentence, model):
    doc = model(sentence)
    for token in doc:
        print(token.text, [t.text for t in token.subtree])
```

12. Let's use this function on the first sentence in our list:

```
print_subtree(sentence, small_model)
```

The result should be as follows. From the subtrees that each word is part of, we can see the grammatical phrases that appear in the sentence, such as the **noun phrase**, any other name, and the **prepositional phrase**, under any other name:

```
I ['I']
have ['have']
seldom ['seldom']
heard ['I', 'have', 'seldom', 'heard', 'him', 'mention', 'her',
'under', 'any', 'other', 'name', '.']
him ['him']
mention ['him', 'mention', 'her', 'under', 'any', 'other',
'name']
her ['her']
under ['under', 'any', 'other', 'name']
any ['any']
other ['other']
name ['any', 'other', 'name']
. ['.']
```

See also

The dependency parse can be visualized graphically using the `displaCy` package, which is part of `spaCy`. Please see *Chapter 87, Visualizing Text Data*, for a detailed recipe on how to do the visualization.

Extracting noun chunks

Noun chunks are known in linguistics as noun phrases. They represent nouns and any words that depend on and accompany nouns. For example, in the sentence *The big red apple fell on the scared cat*, the noun chunks are *the big red apple* and *the scared cat*. Extracting these noun chunks is instrumental to many other downstream NLP tasks, such as named entity recognition and processing entities and relations between them. In this recipe, we will explore how to extract named entities from a text.

Getting ready

We will use the `spaCy` package, which has a function for extracting noun chunks, and the text from the `sherlock_holmes_1.txt` file as an example.

How to do it...

Use the following steps to get the noun chunks from a text:

1. Run the file and language utility notebooks:

    ```
    %run -i "../util/file_utils.ipynb"
    %run -i "../util/lang_utils.ipynb"
    ```

2. Define the function that will print out the noun chunks. The noun chunks are contained in the `doc.noun_chunks` class variable:

    ```
    def print_noun_chunks(text, model):
        doc = model(text)
        for noun_chunk in doc.noun_chunks:
            print(noun_chunk.text)
    ```

3. Read the text from the `sherlock_holmes_1.txt` file and use the function on the resulting text:

    ```
    sherlock_holmes_part_of_text = read_text_file("../data/sherlock_
    holmes_1.txt")
    print_noun_chunks(sherlock_holmes_part_of_text, small_model)
    ```

This is the partial result. See the output of the notebook at `https://github.com/ PacktPublishing/Python-Natural-Language-Processing-Cookbook- Second-Edition/blob/main/Chapter02/noun_chunks_2.3.ipynb` for the full printout. The function gets the pronouns, nouns, and noun phrases that are in the text correctly:

```
Sherlock Holmes
she
the_ woman
I
him
her
any other name
his eyes
she
the whole
…
```

There's more...

Noun chunks are `spaCy` Span objects and have all their properties. See the official documentation at `https://spacy.io/api/token`.

Let's explore some properties of noun chunks:

1. We will define a function that will print out the different properties of noun chunks. It will print the text of the noun chunk, its start and end indices within the `Doc` object, the sentence it belongs to (useful when there is more than one sentence), the root of the noun chunk (its main word), and the chunk's similarity to the word `emotions`. Finally, it will print out the similarity of the whole input sentence to `emotions`:

```python
def explore_properties(sentence, model):
    doc = model(sentence)
    other_span = "emotions"
    other_doc = model(other_span)
    for noun_chunk in doc.noun_chunks:
        print(noun_chunk.text)
        print("Noun chunk start and end", "\t",
            noun_chunk.start, "\t", noun_chunk.end)
        print("Noun chunk sentence:", noun_chunk.sent)
        print("Noun chunk root:", noun_chunk.root.text)
        print(f"Noun chunk similarity to '{other_span}'",
            noun_chunk.similarity(other_doc))
    print(f"Similarity of the sentence '{sentence}' to
        '{other_span}':",
        doc.similarity(other_doc))
```

2. Set the sentence to `All emotions, and that one particularly, were abhorrent to his cold, precise but admirably balanced mind`:

```
sentence = "All emotions, and that one particularly, were
abhorrent to his cold, precise but admirably balanced mind."
```

3. Use the `explore_properties` function on the sentence using the small model:

```
explore_properties(sentence, small_model)
```

This is the result:

```
All emotions
Noun chunk start and end      0    2
Noun chunk sentence: All emotions, and that one particularly,
were abhorrent to his cold, precise but admirably balanced mind.
Noun chunk root: emotions
Noun chunk similarity to 'emotions' 0.4026421588260174
his cold, precise but admirably balanced mind
Noun chunk start and end      11    19
Noun chunk sentence: All emotions, and that one particularly,
were abhorrent to his cold, precise but admirably balanced mind.
Noun chunk root: mind
Noun chunk similarity to 'emotions' -0.036891259527462
Similarity of the sentence 'All emotions, and that one
particularly, were abhorrent to his cold, precise but admirably
balanced mind.' to 'emotions': 0.03174900767577446
```

You will also see a warning message similar to this one due to the fact that the small model does not ship with word vectors of its own:

```
/tmp/ipykernel_1807/2430050149.py:10: UserWarning: [W007] The
model you're using has no word vectors loaded, so the result of
the Span.similarity method will be based on the tagger, parser
and NER, which may not give useful similarity judgements. This
may happen if you're using one of the small models, e.g. `en_
core_web_sm`, which don't ship with word vectors and only use
context-sensitive tensors. You can always add your own word
vectors, or use one of the larger models instead if available.
  print(f"Noun chunk similarity to '{other_span}'", noun_chunk.
similarity(other_doc))
```

4. Now, let's apply the same function to the same sentence with the large model:

```
sentence = "All emotions, and that one particularly, were
abhorrent to his cold, precise but admirably balanced mind."
explore_properties(sentence, large_model)
```

The large model does come with its own word vectors and does not result in a warning:

```
All emotions
Noun chunk start and end    0    2
Noun chunk sentence: All emotions, and that one particularly,
were abhorrent to his cold, precise but admirably balanced mind.
Noun chunk root: emotions
Noun chunk similarity to 'emotions' 0.6302678068015664
his cold, precise but admirably balanced mind
Noun chunk start and end    11    19
Noun chunk sentence: All emotions, and that one particularly,
were abhorrent to his cold, precise but admirably balanced mind.
Noun chunk root: mind
Noun chunk similarity to 'emotions' 0.5744456705692561
Similarity of the sentence 'All emotions, and that one
particularly, were abhorrent to his cold, precise but admirably
balanced mind.' to 'emotions': 0.640366414527618
```

We see that the similarity of the `All emotions` noun chunk is high in relation to the word `emotions`, as compared to the similarity of the `his cold, precise but admirably balanced mind` noun chunk.

> **Important note**
> A larger `spaCy` model, such as `en_core_web_lg`, takes up more space but is more precise.

See also

The topic of semantic similarity will be explored in more detail in *Chapter 3*.

Extracting subjects and objects of the sentence

Sometimes, we might need to find the subject and direct objects of the sentence, and that is easily accomplished with the `spaCy` package.

Getting ready

We will be using the dependency tags from `spaCy` to find subjects and objects. The code uses the `spaCy` engine to parse the sentence. Then, the subject function loops through the tokens, and if the dependency tag contains `subj`, it returns that token's subtree, a `Span` object. There are different subject tags, including `nsubj` for regular subjects and `nsubjpass` for subjects of passive sentences, thus we want to look for both.

How to do it...

We will use the `subtree` attribute of tokens to find the complete noun chunk that is the subject or direct object of the verb (see the *Getting the dependency parse* recipe). We will define functions to find the subject, direct object, dative phrase, and prepositional phrases:

1. Run the file and language utility notebooks:

```
%run -i "../util/file_utils.ipynb"
%run -i "../util/lang_utils.ipynb"
```

2. We will use two functions to find the subject and the direct object of the sentence. These functions will loop through the tokens and return the subtree that contains the token with `subj` or `dobj` in the dependency tag, respectively. Here is the subject function. It looks for the token that has a dependency tag that contains `subj` and then returns the subtree that contains that token. There are several subject dependency tags, including `nsubj` and `nsubjpass` (for the subject of a passive sentence), so we look for the most general pattern:

```
def get_subject_phrase(doc):
    for token in doc:
        if ("subj" in token.dep_):
            subtree = list(token.subtree)
            start = subtree[0].i
            end = subtree[-1].i + 1
            return doc[start:end]
```

3. Here is the direct object function. It works similarly to `get_subject_phrase` but looks for the `dobj` dependency tag instead of a tag that contains `subj`. If the sentence does not have a direct object, it will return `None`:

```
def get_object_phrase(doc):
    for token in doc:
        if ("dobj" in token.dep_):
            subtree = list(token.subtree)
            start = subtree[0].i
            end = subtree[-1].i + 1
            return doc[start:end]
```

4. Assign a list of sentences to a variable, loop through them, and use the preceding functions to print out their subjects and objects:

```
sentences = [
    "The big black cat stared at the small dog.",
    "Jane watched her brother in the evenings.",
    "Laura gave Sam a very interesting book."
]
```

```
for sentence in sentences:
    doc = small_model(sentence)
    subject_phrase = get_subject_phrase(doc)
    object_phrase = get_object_phrase(doc)
    print(sentence)
    print("\tSubject:", subject_phrase)
    print("\tDirect object:", object_phrase)
```

The result will be as follows. Since the first sentence does not have a direct object, None is printed out. For the sentence The big black cat stared at the small dog, the subject is the big black cat and there is no direct object (the small dog is the object of the preposition at). For the sentence Jane watched her brother in the evenings, the subject is Jane and the direct object is her brother. In the sentence Laura gave Sam a very interesting book, the subject is Laura and the direct object is a very interesting book:

```
The big black cat stared at the small dog.
    Subject: The big black cat
    Direct object: None
Jane watched her brother in the evenings.
    Subject: Jane
    Direct object: her brother
Laura gave Sam a very interesting book.
    Subject: Laura
    Direct object: a very interesting book
```

There's more...

We can look for other objects, for example, the dative objects of verbs such as *give* and objects of prepositional phrases. The functions will look very similar, with the main difference being the dependency tags: dative for the dative object function, and pobj for the prepositional object function. The prepositional object function will return a list since there can be more than one prepositional phrase in a sentence:

1. The dative object function checks the tokens for the dative tag. It returns None if there are no dative objects:

```
def get_dative_phrase(doc):
    for token in doc:
        if ("dative" in token.dep_):
            subtree = list(token.subtree)
            start = subtree[0].i
            end = subtree[-1].i + 1
            return doc[start:end]
```

2. We can also combine the subject, object, and dative functions into one with an argument that specifies which object to look for:

```
def get_phrase(doc, phrase):
    # phrase is one of "subj", "obj", "dative"
    for token in doc:
        if (phrase in token.dep_):
            subtree = list(token.subtree)
            start = subtree[0].i
            end = subtree[-1].i + 1
            return doc[start:end]
```

3. Let us now define a sentence with a dative object and run the function for all three types of phrases:

```
sentence = "Laura gave Sam a very interesting book."
doc = small_model(sentence)
subject_phrase = get_phrase(doc, "subj")
object_phrase = get_phrase(doc, "obj")
dative_phrase = get_phrase(doc, "dative")
print(sentence)
print("\tSubject:", subject_phrase)
print("\tDirect object:", object_phrase)
print("\tDative object:", dative_phrase)
```

The result will be as follows. The dative object is Sam:

```
Laura gave Sam a very interesting book.
    Subject: Laura
    Direct object: a very interesting book
    Dative object: Sam
```

4. Here is the prepositional object function. It returns a list of objects of prepositions, which will be empty if there are none:

```
def get_prepositional_phrase_objs(doc):
    prep_spans = []
    for token in doc:
        if ("pobj" in token.dep_):
            subtree = list(token.subtree)
            start = subtree[0].i
            end = subtree[-1].i + 1
            prep_spans.append(doc[start:end])
    return prep_spans
```

5.   Let's define a list of sentences and run the two functions on them:

```
sentences = [
    "The big black cat stared at the small dog.",
    "Jane watched her brother in the evenings."
]
for sentence in sentences:
    doc = small_model(sentence)
    subject_phrase = get_phrase(doc, "subj")
    object_phrase = get_phrase(doc, "obj")
    dative_phrase = get_phrase(doc, "dative")
    prepositional_phrase_objs = \
        get_prepositional_phrase_objs(doc)
    print(sentence)
    print("\tSubject:", subject_phrase)
    print("\tDirect object:", object_phrase)
    print("\tPrepositional phrases:", prepositional_phrase_objs)
```

The result will be as follows:

```
The big black cat stared at the small dog.
  Subject: The big black cat
  Direct object: the small dog
  Prepositional phrases: [the small dog]

Jane watched her brother in the evenings.
  Subject: Jane
  Direct object: her brother
  Prepositional phrases: [the evenings]
```

There is one prepositional phrase in each sentence. In the sentence The big black cat stared at the small dog, it is at the small dog, and in the sentence Jane watched her brother in the evenings, it is in the evenings.

It is left as an exercise for you to find the actual prepositional phrases with prepositions intact instead of just the noun phrases that are dependent on these prepositions.

Finding patterns in text using grammatical information

In this section, we will use the spaCy Matcher object to find patterns in the text. We will use the grammatical properties of the words to create these patterns. For example, we might be looking for verb phrases instead of noun phrases. We can specify grammatical patterns to match verb phrases.

Getting ready

We will be using the spaCy Matcher object to specify and find patterns. It can match different properties, not just grammatical. You can find out more in the documentation at https://spacy. io/usage/rule-based-matching/.

How to do it...

Your steps should be formatted like so:

1. Run the file and language utility notebooks:

    ```
    %run -i "../util/file_utils.ipynb"
    %run -i "../util/lang_utils.ipynb"
    ```

2. Import the Matcher object and initialize it. We need to put in the vocabulary object, which is the same as the vocabulary of the model we will be using to process the text:

    ```
    from spacy.matcher import Matcher
    matcher = Matcher(small_model.vocab)
    ```

3. Create a list of patterns and add them to the matcher. Each pattern is a list of dictionaries, where each dictionary describes a token. In our patterns, we only specify the part of speech for each token. We then add these patterns to the Matcher object. The patterns we will be using are a verb by itself (for example, *paints*), an auxiliary followed by a verb (for example, was observing), an auxiliary followed by an adjective (for example, were late), and an auxiliary followed by a verb and a preposition (for example, were staring at). This is not an exhaustive list; feel free to come up with other examples:

    ```
    patterns = [
        [{"POS": "VERB"}],
        [{"POS": "AUX"}, {"POS": "VERB"}],
        [{"POS": "AUX"}, {"POS": "ADJ"}],
        [{"POS": "AUX"}, {"POS": "VERB"}, {"POS": "ADP"}]
    ]
    matcher.add("Verb", patterns)
    ```

4. Read in the small part of the *Sherlock Holmes* text and process it using the small model:

    ```
    sherlock_holmes_part_of_text = read_text_file("../data/sherlock_
    holmes_1.txt")
    doc = small_model(sherlock_holmes_part_of_text)
    ```

5. Now, we find the matches using the `Matcher` object and the processed text. We then loop through the matches and print out the match ID, the string ID (the identifier of the pattern), the start and end of the match, and the text of the match:

```
matches = matcher(doc)
for match_id, start, end in matches:
    string_id = small_model.vocab.strings[match_id]
    span = doc[start:end]
    print(match_id, string_id, start, end, span.text)
```

The result will be as follows:

```
14677086776663181681 Verb 14 15 heard
14677086776663181681 Verb 17 18 mention
14677086776663181681 Verb 28 29 eclipses
14677086776663181681 Verb 31 32 predominates
14677086776663181681 Verb 43 44 felt
14677086776663181681 Verb 49 50 love
14677086776663181681 Verb 63 65 were abhorrent
14677086776663181681 Verb 80 81 take
14677086776663181681 Verb 88 89 observing
14677086776663181681 Verb 94 96 has seen
14677086776663181681 Verb 95 96 seen
14677086776663181681 Verb 103 105 have placed
14677086776663181681 Verb 104 105 placed
14677086776663181681 Verb 114 115 spoke
14677086776663181681 Verb 120 121 save
14677086776663181681 Verb 130 132 were admirable
14677086776663181681 Verb 140 141 drawing
14677086776663181681 Verb 153 154 trained
14677086776663181681 Verb 157 158 admit
14677086776663181681 Verb 167 168 adjusted
14677086776663181681 Verb 171 172 introduce
14677086776663181681 Verb 173 174 distracting
14677086776663181681 Verb 178 179 throw
14677086776663181681 Verb 228 229 was
```

The code finds some of the verb phrases in the text. Sometimes, it finds a partial match that is part of another match. Weeding out these partial matches is left as an exercise.

See also

We can use other attributes apart from parts of speech. It is possible to match on the text itself, its length, whether it is alphanumeric, the punctuation, the word's case, the `dep_` and `morph` attributes, lemma, entity type, and others. It is also possible to use regular expressions on the patterns. For more information, see the spaCy documentation: `https://spacy.io/usage/rule-based-matching`.

3

Representing Text – Capturing Semantics

Representing the meaning of words, phrases, and sentences in a form that's understandable to computers is one of the pillars of NLP processing. Machine learning, for example, represents each data point as a list of numbers (a fixed-size vector), and we are faced with the question of how to turn words and sentences into these vectors. Most NLP tasks start by representing the text in some numeric form, and in this chapter, we show several ways to do that.

First, we will create a simple classifier to demonstrate the effectiveness of each method of encoding, and then we will use it to test the different encoding methods. We will also learn how to turn phrases such as *fried chicken* into vectors – that is, how to train a `word2vec` model for phrases. Finally, we will see how to use vector-based search.

For a theoretical background on some of the concepts discussed in this section, refer to *Building Machine Learning Systems with Python* by Coelho et al. This book will explain the basics of building a machine learning project, such as training and test sets, as well as metrics used to evaluate such projects, including precision, recall, F1, and accuracy.

Here are the recipes that are covered in this chapter:

- Creating a simple classifier
- Putting documents into a bag of words
- Constructing an *N*-gram model
- Representing texts with TF-IDF
- Using word embeddings
- Training your own embeddings model
- Using BERT and OpenAI embeddings instead of word embeddings
- Using **retrieval augmented generation** (**RAG**)

Technical requirements

The code for this chapter is located at `https://github.com/PacktPublishing/Python-Natural-Language-Processing-Cookbook-Second-Edition/tree/main/Chapter03`. Packages that are required for this chapter should be installed automatically via the `poetry` environment.

In addition, we will use models and datasets located at the following URLs. The Google `word2vec` model is a model that represents words as vectors, and the IMDB dataset contains movie titles, genres, and descriptions. Download them into the `data` folder inside the `root` directory:

- **The Google `word2vec` model**: `https://drive.google.com/file/d/0B7XkCwpI5KDYNlNUTTlSS21pQmM/edit?resourcekey=0-wjGZdNAUop6WykTtMip30g`
- **The IMDB movie dataset**: `https://github.com/venusanvi/imdb-movies/blob/main/IMDB-Movie-Data.csv` (also available in the book's GitHub repo)

In addition to the preceding files, we will use various functions from a simple classifier that we will create in the first recipe. This file is available at `https://github.com/PacktPublishing/Python-Natural-Language-Processing-Cookbook-Second-Edition/blob/main/util/util_simple_classifier.ipynb`.

Creating a simple classifier

The reason why we need to represent text as vectors is to make it into a computer-readable form. Computers can't understand words but are good at manipulating numbers. One of the main NLP tasks is the classification of texts, and we are going to create a classifier for movie reviews. We will use the same classifier code but with different methods of creating vectors from text.

In this section, we will create the classifier that will assign either negative or positive sentiment to *Rotten Tomatoes* reviews, a dataset available through Hugging Face, a large repository of open source models and datasets. We will then use a baseline method, where we encode the text by counting the number of different parts of speech present in it (verbs, nouns, proper nouns, adjectives, adverbs, auxiliary verbs, pronouns, numbers, and punctuation).

By the end of this recipe, we will have created a separate file with functions that create the dataset and train the classifier. We will use this file throughout the chapter to test different encoding methods.

Getting ready

In this recipe, we will create a simple classifier for movie reviews. It will be a **logistic regression classifier** from the `sklearn` package.

How to do it...

We will load the Rotten Tomatoes dataset from Hugging Face. We will use just part of the dataset so that the training time is not very long:

1. Import the file and language `util` notebooks:

    ```
    %run -i "../util/file_utils.ipynb"
    %run -i "../util/lang_utils.ipynb"
    ```

2. Load the train and test datasets from Hugging Face (the `datasets` package). For both the training and test sets, we will select the first and last 15% of the data instead of loading the full datasets. The full dataset is large, and it takes a long time to train the model:

    ```
    from datasets import load_dataset
    train_dataset = load_dataset("rotten_tomatoes",
        split="train[:15%]+train[-15%:]")
    test_dataset = load_dataset("rotten_tomatoes",
        split="test[:15%]+test[-15%:]")
    ```

3. Print out the length of each dataset:

    ```
    print(len(train_dataset))
    print(len(test_dataset))
    ```

    The output should be as follows:

    ```
    2560
    320
    ```

4. Here, we create the `POS_vectorizer` class. This class has a method, `vectorize`, that processes the text and counts the number of verbs, nouns, proper nouns, adjectives, adverbs, auxiliary verbs, pronouns, numbers, and punctuation marks. The class needs a `spaCy` model to process the text. Each piece of text is turned into a vector of size 10. The first element of the vector is the length of the text, and the other numbers indicate the number of words in the text of that particular part of speech:

    ```
    class POS_vectorizer:
        def __init__(self, spacy_model):
            self.model = spacy_model

        def vectorize(self, input_text):
            doc = self.model(input_text)
            vector = []
            vector.append(len(doc))
            pos = {"VERB":0, "NOUN":0, "PROPN":0, "ADJ":0,
                "ADV":0, "AUX":0, "PRON":0, "NUM":0, "PUNCT":0}
            for token in doc:
    ```

```
            if token.pos_ in pos:
                pos[token.pos_] += 1
        vector_values = list(pos.values())
        vector = vector + vector_values
        return vector
```

5. Now, we can test out the `POS_vectorizer` class. We take the first review's text to process and create the vectorizer using the small spaCy model. We then vectorize the text using the newly created class:

```
sample_text = train_dataset[0]["text"]
vectorizer = POS_vectorizer(small_model)
vector = vectorizer.vectorize(sample_text)
```

6. Let's print the text and the vector:

```
print(sample_text)
print(vector)
```

The result should look like this. We can see that the vector correctly counts the parts of speech. For example, there are five punctuation marks (two quotes, one comma, one dot, and one dash):

```
the rock is destined to be the 21st century's new " conan "
and that he's going to make a splash even greater than arnold
schwarzenegger , jean-claud van damme or steven segal .
[38, 3, 8, 3, 4, 1, 3, 1, 0, 5]
```

7. We will now prepare the data for training our classifier. We first import the `pandas` and the `numpy` packages and then create two dataframes, one for training and the other one for testing. In each dataset, we create a new column called `vector` that contains the vector for this piece of text. We use the `apply` method to turn the text into vectors and store them in the new column. In this method, we pass in a lambda function that takes a piece of text and applies the `vectorize` method of the `POS_vectorizer` class to that piece of text. We then turn the vector and the label columns into `numpy` arrays to have the data in the right format for the classifier. We use the `np.stack` method for the vector, since it's already a list, and the `to_numpy` method for the review labels, since they are just numbers:

```
import pandas as pd
import numpy as np
train_df = train_dataset.to_pandas()
train_df.sample(frac=1)
test_df = test_dataset.to_pandas()
train_df["vector"] = train_df["text"].apply(
    lambda x: vectorizer.vectorize(x))
test_df["vector"] = test_df["text"].apply(
    lambda x: vectorizer.vectorize(x))
X_train = np.stack(train_df["vector"].values, axis=0)
X_test = np.stack(test_df["vector"].values, axis=0)
```

```
y_train = train_df["label"].to_numpy()
y_test = test_df["label"].to_numpy()
```

8. Now, we will train the classifier. We will choose the logistic regression algorithm, since it is one of the simplest algorithms, as well as one of the fastest. First, we import the `LogisticRegression` class and the `classification_report` methods from `sklearn`. Then, we create the `LogisticRegression` object, and finally, train it on the data from the previous step:

```
from sklearn.linear_model import LogisticRegression
from sklearn.metrics import classification_report
clf = LogisticRegression(C=0.1)
clf = clf.fit(X_train, y_train)
```

9. We can test the classifier on the test data by applying the `predict` method to the vectors in the test data and print out the classification report. We can see that the overall accuracy is low, slightly above chance. This is because the vector representation we used is very crude. We will use other vectors in the next sections and see how they affect the classifier results:

```
test_df["prediction"] = test_df["vector"].apply(
    lambda x: clf.predict([x])[0])
print(classification_report(test_df["label"],
    test_df["prediction"]))
```

The output should be similar to this:

```
              precision    recall  f1-score   support

           0       0.59      0.54      0.56       160
           1       0.57      0.62      0.60       160

    accuracy                           0.58       320
   macro avg       0.58      0.58      0.58       320
weighted avg       0.58      0.58      0.58       320
```

There's more…

We will now turn the preceding code into several functions so that we can only vary the vectorizer being used in the construction of the dataset. The resulting file is located at https://github. com/PacktPublishing/Python-Natural-Language-Processing-Cookbook-Second-Edition/blob/main/util/util_simple_classifier.ipynb. The resulting code will look as follows:

1. Import the necessary packages:

```
from datasets import load_dataset
import pandas as pd
```

```
import numpy as np
from sklearn.linear_model import LogisticRegression
from sklearn.metrics import classification_report
```

2. Define the function that will create and return the training and test dataframes. It will create them from the Rotten Tomatoes dataset from Hugging Face:

```
def load_train_test_dataset_pd():
    train_dataset = load_dataset("rotten_tomatoes",
        split="train[:15%]+train[-15%:]")
    test_dataset = load_dataset("rotten_tomatoes",
        split="test[:15%]+test[-15%:]")
    train_df = train_dataset.to_pandas()
    train_df.sample(frac=1)
    test_df = test_dataset.to_pandas()
    return (train_df, test_df)
```

3. This function takes the dataframes and the `vectorize` method and creates the numpy arrays for the training and test data. This will allow us to train the logistic regression classifier using the created vectors:

```
def create_train_test_data(train_df, test_df, vectorize):
    train_df["vector"] = train_df["text"].apply(
        lambda x: vectorize(x))
    test_df["vector"] = test_df["text"].apply(
        lambda x: vectorize(x))
    X_train = np.stack(train_df["vector"].values, axis=0)
    X_test = np.stack(test_df["vector"].values, axis=0)
    y_train = train_df["label"].to_numpy()
    y_test = test_df["label"].to_numpy()
    return (X_train, X_test, y_train, y_test)
```

4. This function trains a logistic regression classifier on the given training data:

```
def train_classifier(X_train, y_train):
    clf = LogisticRegression(C=0.1)
    clf = clf.fit(X_train, y_train)
    return clf
```

5. This final function takes in the test data and the trained classifier and prints out the classification report:

```
def test_classifier(test_df, clf):
    test_df["prediction"] = test_df["vector"].apply(
        lambda x: clf.predict([x])[0])
    print(classification_report(test_df["label"],
        test_df["prediction"]))
```

In each succeeding section that demonstrates a new vectorizing method, we will use this file to pre-load the necessary functions to test the classification result. This will allow us to evaluate the different vectorizing methods. We will only vary the vectorizer while keeping the classifier the same. When the classifier performs better, it reflects how well the underlying vectorizer represents the text.

Putting documents into a bag of words

A **bag of words** is the simplest way of representing a text. We treat our text as a collection of *documents*, where documents are anything from sentences to scientific articles to blog posts or whole books. Since we usually compare different documents to each other or use them in a larger context of other documents, we work with a collection of documents, not just a single document.

The bag of words method uses a "training" text that provides it with a list of words that it should consider. When encoding new sentences, it counts the number of occurrences each word makes in the document, and the final vector includes those counts for each word in the vocabulary. This representation can then be fed into a machine learning algorithm.

The reason this vectorizing method is called a *bag of words* is that it does not take into account the relationships of words between themselves and only counts the number of occurrences of each word. The decision on what represents a document lies with the engineer and, in many cases, will be obvious. For example, if you are working on classifying tweets that belong to a particular topic, a single tweet will be your document. If, conversely, you would like to find out which chapters of a book are most similar to a book you already have, then chapters are documents.

In this recipe, we will create a bag of words for the Rotten Tomatoes reviews. Our documents will be the reviews. We then test the encoding using a bag of words by building a logistic regression classifier, using code from the previous recipe.

Getting ready

For this recipe, we will use the CountVectorizer class from the sklearn package. It is included in the poetry environment. The CountVectorizer class is specifically designed to count the number of occurrences of each word in a text.

How to do it...

Our code will take a set of documents – in this case, reviews – and represent them as a matrix of vectors. We will use the Rotten Tomatoes reviews dataset from Hugging Face for this task:

1. Run the simple classifier utility file, and then import the CountVectorizer object and the sys package. We will need the sys package to change the printing options:

```
%run -i "../util/util_simple_classifier.ipynb"
from sklearn.feature_extraction.text import CountVectorizer
import sys
```

2. Load the training and testing dataframes by using the function from the `util_simple_classifier.ipynb` file. We created this function in the previous recipe, *Creating a simple classifier*. The function loads 15% of the Rotten Tomatoes dataset into a `pandas` dataframe and randomizes its order. It might take a few minutes to run:

```
(train_df, test_df) = load_train_test_dataset_pd()
```

3. Create the vectorizer, fit it on the training data, and print out the resulting matrix. We will use the `max_df` parameter to specify which words should be used as stop words. In this case, we specify that words that appear in more than 40% of the documents should be ignored when constructing the vectorizer. You should experiment and see exactly which value of `max_df` would suit your use case. We then fit the vectorizer on the `text` column of the `train_df` dataframe:

```
vectorizer = CountVectorizer(max_df=0.4)
X = vectorizer.fit_transform(train_df["text"])
print(X)
```

The resulting matrix is a `scipy.sparse._csr.csr_matrix` object, and the beginning of its printout looks like this. The format of a sparse matrix is `(row, column) value`. In our case, this means (the document index, word index) followed by the frequency. In our example, the first review, which is the first document, is document number 0, and it contains words with indices `6578`, `4219`, and others. The frequencies of these words are 1 and 2, respectively.

```
(0, 6578)    1
(0, 4219)    1
(0, 2106)    1
(0, 8000)    2
(0, 717)    1
(0, 42)    1
(0, 1280)    1
(0, 5260)    1
(0, 1607)    1
(0, 7889)    1
(0, 3630)    1
...
```

4. In most cases, we use a different format to represent vectors, a dense matrix that is easier to use in practice. Instead of specifying rows and columns with numbers, they are inferred from the position of the value. We will now create a dense matrix and print it:

```
dense_matrix = X.todense()
print(dense_matrix)
```

The resulting matrix is a NumPy matrix object, where each review is a vector. You see that most values in the matrix are zeroes, as expected, since each review only uses a handful of words, while the vector collects counts for each word in the vocabulary, or each word in all of the reviews. Any words that are not in the vectorizer's vocabulary will not be included in the vector:

```
[[0 0 0 ... 0 0 0]
 [0 0 0 ... 0 0 0]
 [0 0 0 ... 0 0 0]
 ...
 [0 0 0 ... 0 0 0]
 [0 0 0 ... 0 0 0]
 [0 0 0 ... 0 0 0]]
```

5. We can see all the words used in the document set and the length of the vocabulary. This can be used as a sanity check and to see whether there are any irregularities in the vocabulary:

```
print(vectorizer.get_feature_names_out())
print(len(vectorizer.get_feature_names_out()))
```

The result will be as follows. If you want to see the full, non-truncated list, use the set_ printoptions function used in *step 8*:

```
['10' '100' '101' ... 'zone' 'ótimo' 'últimos']
8856
```

6. We can also see all the stop words used by the vectorizer:

```
print(vectorizer.stop_words_)
```

The result is three words, and, the, and of, that appear in more than 40% of reviews:

```
{'and', 'the', 'of'}
```

7. We can now also use the CountVectorizer object to represent new reviews that were not in the original document set. This is done when we have a trained model and want to test it on new, unseen samples. We will use the first review in the test dataset. To get the first review in the test set, we will use the pandas iat function.

```
first_review = test_df['text'].iat[0]
print(first_review)
```

The first review looks as follows:

```
lovingly photographed in the manner of a golden book sprung
to life , stuart little 2 manages sweetness largely without
stickiness .
```

8. Now, we will create both a sparse and a dense vector from the first review. The transform method of the vectorizer expects a list of strings, so we will create a list. We also set the print option to print out the whole vector instead of just part of it:

```
sparse_vector = vectorizer.transform([first_review])
print(sparse_vector)
dense_vector = sparse_vector.todense()
```

```
np.set_printoptions(threshold=sys.maxsize)
print(dense_vector)
np.set_printoptions(threshold=False)
```

The dense vector is very long and is mostly zeroes, as expected:

```
  (0,  955)    1
  (0,  3968)   1
  (0,  4451)   1
  (0,  4562)   1
  (0,  4622)   1
  (0,  4688)   1
  (0,  4779)   1
  (0,  4792)   1
  (0,  5764)   1
  (0,  7547)   1
  (0,  7715)   1
  (0,  8000)   1
  (0,  8734)   1
[[0 0 0 0 0 0 0 0 0 0 0 0 0 0 0 0 0 0 0 0 0 0 0 0 0 0 0 0 0 0 0 0 0 0
  0 0 0 0 0
  0 0 0 0 0 0 0 0 0 0 0 0 0 0 0 0 0 0 0 0 0 0 0 0 0 0 0 0 0 0 0 0 0 0 0
  0 0 0 0 0
...]]
```

9. We can use a different method to calculate stop words. Here, stop words are calculated by setting an absolute threshold on the word frequency. In this case, we use all words whose frequency is lower than 300 across documents. You can see that the stop-word list is now larger.

```
vectorizer = CountVectorizer(max_df=300)
X = vectorizer.fit_transform(train_df["text"])
print(vectorizer.stop_words_)
```

The result will be as follows:

```
{'but', 'this', 'its', 'as', 'to', 'and', 'the', 'is', 'film',
'for', 'it', 'an', 'of', 'that', 'movie', 'with', 'in'}
```

10. Finally, we can provide our own list of stop words to the vectorizer. These words will be ignored by the vectorizer and not represented in the vector. This is useful if you have very specific words you would like to ignore:

```
vectorizer = CountVectorizer(stop_words=['the', 'this',
    'these', 'in', 'at', 'for'])
X = vectorizer.fit_transform(train_df["text"])
```

11. We will now test the effect of this bag-of-words vectorizer on the simple classifier, using the functions we defined in the previous recipe. First, we create the vectorizer, specifying to use only words that appear in less than 80% of the documents. Then, we load the training and test dataframes. We fit the vectorizer on the training set reviews. We create a vectorize function using the vectorizer and pass it on to the `create_train_test_data` function, along with the training and test data frames. We then train the classifier and test it on the testing data. We can see that this vectorizing method gives us much better results than the simple part-of-speech count vector we used in the previous section:

```
vectorizer = CountVectorizer(max_df=0.8)
(train_df, test_df) = load_train_test_dataset_pd()
X = vectorizer.fit_transform(train_df["text"])
vectorize = lambda x: vectorizer.transform([x]).toarray()[0]
(X_train, X_test, y_train, y_test) = create_train_test_data(
    train_df, test_df, vectorize)
clf = train_classifier(X_train, y_train)
test_classifier(test_df, clf)
```

The result will be similar to this:

	precision	recall	f1-score	support
0	0.74	0.72	0.73	160
1	0.73	0.75	0.74	160
accuracy			0.74	320
macro avg	0.74	0.74	0.74	320
weighted avg	0.74	0.74	0.74	320

Constructing an N-gram model

Representing a document as a bag of words is useful, but semantics is about more than just words in isolation. To capture word combinations, an **n-gram model** is useful. Its vocabulary consists of not just words but also word sequences, or *n*-grams.

We will build a **bigram model** in this recipe, where bigrams are sequences of two words.

Getting ready

The `CountVectorizer` class is very versatile and allows us to construct *n*-gram models. We will use it in this recipe and test it with a simple classifier.

In this recipe, I make comparisons of the code and its results to the ones in the *Putting documents into a bag of words* recipe, since the two are very similar, but they have a few differing characteristics.

How to do it...

1. Run the simple classifier notebook and import the `CountVectorizer` class:

```
%run -i "../util/util_simple_classifier.ipynb"
from sklearn.feature_extraction.text import CountVectorizer
```

2. Create the training and test dataframes using code from the `util_simple_classifier.ipynb` notebook:

```
(train_df, test_df) = load_train_test_dataset_pd()
```

3. Create a new vectorizer class. In this case, we will use the `ngram_range` argument. The `CountVectorizer` class, when the `ngram_range` argument is set, counts not only individual words but also word combinations, where the number of words in the combinations depends on the numbers provided to the `ngram_range` argument. We provided `ngram_range=(1,2)` as the argument, which means that the number of words in the combinations ranges from 1 to 2, so unigrams and bigrams are counted:

```
bigram_vectorizer = CountVectorizer(
    ngram_range=(1, 2), max_df=0.8)
X = bigram_vectorizer.fit_transform(train_df["text"])
```

4. Print the vocabulary of the vectorizer and its length. As you can see, the length of the vocabulary is much larger than the length of the unigram vectorizer, since we use two-word combinations in addition to single words:

```
print(bigram_vectorizer.get_feature_names_out())
print(len(bigram_vectorizer.get_feature_names_out()))
```

The result should look like this:

```
['10' '10 inch' '10 set' ... 'ótimo esforço' 'últimos' 'últimos
tiempos']
40552
```

5. Now, we take the first review in the testing dataframe and get its dense vector. The result looks very similar to the vector output in the *Putting documents into a bag of words* recipe, with the only difference that now the output is longer, as it includes not just individual words but also bigrams, or sequences of two words:

```
first_review = test_df['text'].iat[0]
dense_vector = bigram_vectorizer.transform(
    [first_review]).todense()
print(dense_vector)
```

The printout looks like this:

```
[[0 0 0 ... 0 0 0]]
```

6. Finally, we train a simple classifier using the new bigram vectorizer. The resulting accuracy is slightly worse than the accuracy of the classifier that uses a unigram vectorizer from the previous section. There are several possible reasons for this. One is that the vectors are now much longer and still mostly zeroes. The other is that we can actually see that not all reviews are in English, so it is hard for the classifier to generalize the incoming data:

```
vectorize = \
    lambda x: bigram_vectorizer.transform([x]).toarray()[0]
(X_train, X_test, y_train, y_test) = create_train_test_data(
    train_df, test_df, vectorize)
clf = train_classifier(X_train, y_train)
test_classifier(test_df, clf)
```

The output will be as follows:

	precision	recall	f1-score	support
0	0.72	0.75	0.73	160
1	0.74	0.71	0.72	160
accuracy			0.73	320
macro avg	0.73	0.73	0.73	320
weighted avg	0.73	0.73	0.73	320

There's more...

We can use trigrams, quadrigrams, and so on in the vectorizer by providing the corresponding tuple to the ngram_range argument. The downside of this is the ever-expanding vocabulary and the growth of sentence vectors, since each sentence vector has to have an entry for each word in the input vocabulary.

It is also possible to represent character n-grams using the CountVectorizer class. In this case, you would count the occurrence of character sequences instead of word sequences.

Representing texts with TF-IDF

We can go one step further and use the TF-IDF algorithm to count words and n-grams in incoming documents. **TF-IDF** stands for **term frequency-inverse document frequency** and gives more weight to words that are unique to a document than to words that are frequent but repeated throughout most documents. This allows us to give more weight to words uniquely characteristic of particular documents.

In this recipe, we will use a different type of vectorizer that can apply the TF-IDF algorithm to the input text and build a small classifier.

Getting ready

We will use the `TfidfVectorizer` class from the `sklearn` package. The features of the `TfidfVectorizer` class should be familiar from the two previous recipes, *Putting documents into a bag of words* and *Constructing an N-gram model*. We will again use the Rotten Tomatoes review dataset from Hugging Face.

How to do it...

Here are the steps to build and use the TF-IDF vectorizer:

1. Run the small classifier notebook and import the `TfidfVectorizer` class:

```
%run -i "../util/util_simple_classifier.ipynb"
from sklearn.feature_extraction.text import TfidfVectorizer
```

2. Create the training and test dataframes using the `load_train_test_dataset_pd()` function:

```
(train_df, test_df) = load_train_test_dataset_pd()
```

3. Create the vectorizer and fit it on the training text. We will use the `max_df` parameter to exclude stop words – in this case, words that are more frequent than 300:

```
vectorizer = TfidfVectorizer(max_df=300)
vectorizer.fit(train_df["text"])
```

4. To make sure the result makes sense, we will print the vectorizer vocabulary and its length. Since we are just using unigrams, the size of the vocabulary should be the same as the one in the bag-of-words recipe:

```
print(vectorizer.get_feature_names_out())
print(len(vectorizer.get_feature_names_out()))
```

The result should be as follows. The length of the vocabulary should be the same as the one we get in the bag-of-words recipe, since we are not using *n*-grams:

```
['10' '100' '101' ... 'zone' 'ótimo' 'últimos']
8842
```

5. Now, let's take the first review in the test dataframe and vectorizer it. We then print the dense vector. To learn more about the difference between sparse and dense vectors, see the *Putting documents into a bag of words* recipe. Note that the values in the vector are now floats and not integers. This is because the individual values are now ratios and not counts:

```
first_review = test_df['text'].iat[0]
dense_vector = vectorizer.transform([first_review]).todense()
print(dense_vector)
```

The result should be as follows:

```
[[0. 0. 0. ... 0. 0. 0.]]
```

6. Now, let's train the classifier. We can see that the scores are slightly higher than those for a bag-of-words classifier, both the unigram and *n*-gram versions:

```
vectorize = lambda x: vectorizer.transform([x]).toarray()[0]
(X_train, X_test, y_train, y_test) = create_train_test_data(
    train_df, test_df, vectorize)
clf = train_classifier(X_train, y_train)
test_classifier(test_df, clf)
```

The printout of the test scores will look similar to this:

	precision	recall	f1-score	support
0	0.76	0.72	0.74	160
1	0.74	0.78	0.76	160
accuracy			0.75	320
macro avg	0.75	0.75	0.75	320
weighted avg	0.75	0.75	0.75	320

How it works...

The `TfidfVectorizer` class works almost exactly like the `CountVectorizer` class, differing only in the way the **word frequencies** are calculated, so most of the steps should be familiar. Word frequencies are calculated as follows. For each word, the overall frequency is a product of the **term frequency** and the **inverse document frequency**. Term frequency is the number of times a word occurs in the document. Inverse document frequency is the total number of documents divided by the number of documents where the word occurs. Usually, these frequencies are logarithmically scaled.

This is done using the following formulas:

$$TF = \frac{Number\ of\ times\ a\ term\ appears\ in\ the\ document}{Total\ number\ of\ words\ in\ the\ document}$$

$$IDF = \frac{Total\ number\ of\ documents}{Number\ of\ documents\ where\ term\ occurs}$$

$$TF - IDF = TF * IDF$$

There's more…

We can build `TfidfVectorizer` and use **character n-grams** instead of **word n-grams**. Character n-grams use the character, not the word, as their basic unit. For example, if we were to build character n-grams for the phrase *the woman* with the n-gram range (1, 3), it would be the [t, h, e, w, o, m, a, n, th, he, wo, om, ma, an, the, wom, oma, man] set. In some experimental settings, models based on character n-grams perform better than word-based n-gram models.

We will use the small Sherlock Holmes text file, `sherlock_holmes_1.txt`, found at https://github.com/PacktPublishing/Python-Natural-Language-Processing-Cookbook-Second-Edition/blob/main/data/sherlock_holmes_1.txt, and the same class, `TfidfVectorizer`. We will not need a tokenizer function or a stop-word list, since the unit of analysis is the character and not the word. The steps to create the vectorizer and analyze a sentence are as follows:

1. Create a new vectorizer object that uses the `char_wb` analyzer, and then fit it on the training text:

```
tfidf_char_vectorizer = TfidfVectorizer(
    analyzer='char_wb', ngram_range=(1,5))
tfidf_char_vectorizer = tfidf_char_vectorizer.fit(
    train_df["text"])
```

2. Print the vectorizer vocabulary and its length:

```
print(list(tfidf_char_vectorizer.get_feature_names_out()))
print(len(tfidf_char_vectorizer.get_feature_names_out()))
```

The partial result will look like this:

```
[' ', ' !', ' ! ', ' "', ' " ', ' $', ' $5', ' $50', ' $50-', '
$9', ' $9 ', ' &', ' & ', " '", " ' ", " ' ", " '5", " '50", " '50'", "
'6", " '60", " '60s", " '7", " '70", " '70'", " '70s", " '[", "
'[h", " '[ho", " 'a", " 'a ", " 'a'", " 'a' ", " 'ab", " 'aba",
" 'ah", " 'ah ", " 'al", " 'alt", " 'an", " 'ana", " 'ar", "
'are", " 'b", " 'ba", " 'bar", " 'be", " 'bee", " 'bes", " 'bl",
" 'blu", " 'br", " 'bra", " 'bu", " 'but", " 'c", " 'ch", "
'cha", " 'co", " 'co-", " 'com", " 'd", " 'di", " 'dif", " 'do",
" 'dog", " 'du", " 'dum", " 'e", " 'ed", " 'edg", " 'em", " 'em
", " 'ep", " 'epi", " 'f", " 'fa", " 'fac", " 'fat", " 'fu", "
'fun", " 'g", " 'ga", " 'gar", " 'gi", " 'gir", " 'gr", " 'gra",
" 'gu", " 'gue", " 'guy", " 'h", " 'ha", " 'hav", " 'ho", "
'hos", " 'how", " 'i", " 'i ", " 'if", " 'if ", " 'in", " 'in ",
" 'is",…]
51270
```

3. Create the `vectorize` method using the new vectorizer, and then create the training and test data. Train the classifier and then test it:

```
vectorize = lambda x: tfidf_char_vectorizer.transform([
    x]).toarray()[0]
```

```
(X_train, X_test, y_train, y_test) = create_train_test_data(
    train_df, test_df, vectorize)
clf = train_classifier(X_train, y_train)
test_classifier(test_df, clf)
```

The result will be similar to this:

```
              precision    recall  f1-score   support

           0       0.74      0.74      0.74       160
           1       0.74      0.74      0.74       160

    accuracy                           0.74       320
   macro avg       0.74      0.74      0.74       320
weighted avg       0.74      0.74      0.74       320
```

See also

- You can find out more about term weighting at https://scikit-learn.org/stable/
  modules/feature_extraction.html#tfidf-term-weighting

- For more information about TfidfVectorizer, see https://scikit-learn.
  org/stable/modules/generated/sklearn.feature_extraction.text.
  TfidfVectorizer.html

Using word embeddings

In this recipe, we will switch gears and learn how to represent *words* using word embeddings, which are powerful because they are a result of training a neural network that predicts a word from all other words in the sentence. Embeddings are also vectors, but usually of a much smaller size, 200 or 300. The resulting vector embeddings are similar for words that occur in similar contexts. Similarity is usually measured by calculating the cosine of the angle between two vectors in the hyperplane, with 200 or 300 dimensions. We will use the embeddings to show these similarities.

Getting ready

In this recipe, we will use a pretrained word2vec model, which can be found at https://github.com/mmihaltz/word2vec-GoogleNews-vectors. Download the model and unzip it in the data directory. You should now have a file with the .../data/GoogleNews-vectors-negative300.bin.gz path.

We will also use the gensim package to load and use the model. It should be installed within the poetry environment.

The notebook is located at `https://github.com/PacktPublishing/Python-Natural-Language-Processing-Cookbook-Second-Edition/blob/main/Chapter03/3.5_word_embeddings.ipynb`.

How to do it...

We will load the model, demonstrate some features of the `gensim` package, and then compute a sentence vector using the word embeddings:

1. Run the simple classifier file:

    ```
    %run -i "../util/simple_classifier.ipynb"
    ```

2. Import the `gensim` package:

    ```
    import gensim
    ```

3. Load the pretrained model. If you get an error at this step, make sure you have downloaded the model in the `data` directory:

    ```
    model = gensim.models.KeyedVectors.load_word2vec_format(
        '../data/GoogleNews-vectors-negative300.bin.gz',
        binary=True)
    ```

4. Using the pretrained model, we can now load individual word vectors. Here, we load the word vector for the word *king*. We have to lowercase it, since all the words in the model are lowercase. The result is a long vector that represents this word in the `word2vec` model:

    ```
    vec_king = model['king']
    print(vec_king)
    ```

    The result will be as follows:

    ```
    [ 1.25976562e-01  2.97851562e-02  8.60595703e-03  1.39648438e-01
     -2.56347656e-02 -3.61328125e-02  1.11816406e-01 -1.98242188e-01
      5.12695312e-02  3.63281250e-01 -2.42187500e-01 -3.02734375e-01
     -1.77734375e-01 -2.49023438e-02 -1.67968750e-01 -1.69921875e-01
      3.46679688e-02  5.21850586e-03  4.63867188e-02  1.28906250e-01
      1.36718750e-01  1.12792969e-01  5.95703125e-02  1.36718750e-01
      1.01074219e-01 -1.76757812e-01 -2.51953125e-01  5.98144531e-02
      3.41796875e-01 -3.11279297e-02  1.04492188e-01  6.17675781e-
    02   ...]
    ```

5. We can also get words that are most similar to a given word. For example, let's print out the words most similar to *apple* and *tomato*. The output prints out the words that are the most similar (i.e., occur in similar contexts) and the similarity score. The score is the cosine distance between a pair of vectors – in this case, representing a pair of words. The larger the score, the more similar the two words. The results make sense, since the words most similar to *apple* are mostly fruits, and the words most similar to *tomato* are mostly vegetables:

```
print(model.most_similar(['apple'], topn=15))
print(model.most_similar(['tomato'], topn=15))
```

The result is as follows:

```
[('apples', 0.720359742641449), ('pear', 0.6450697183609009),
('fruit', 0.6410146355628967), ('berry', 0.6302295327186584),
('pears', 0.613396167755127), ('strawberry',
0.6058260798454285), ('peach', 0.6025872826576233), ('potato',
0.5960935354232788), ('grape', 0.5935863852500916),
('blueberry', 0.5866668224334717), ('cherries',
0.5784382820129395), ('mango', 0.5751855969429016), ('apricot',
0.5727777481079102), ('melon', 0.5719985365867615), ('almond',
0.5704829692840576)]
[('tomatoes', 0.8442263007164001), ('lettuce',
0.7069936990737915), ('asparagus', 0.7050934433937073),
('peaches', 0.6938520669937134), ('cherry_tomatoes',
0.6897529363632202), ('strawberry', 0.6888598799705505),
('strawberries', 0.6832595467567444), ('bell_peppers',
0.6813562512397766), ('potato', 0.6784172058105469),
('cantaloupe', 0.6780219078063965), ('celery',
0.675195574760437), ('onion', 0.6740139722824097), ('cucumbers',
0.6706333160400391), ('spinach', 0.6682621240615845),
('cauliflower', 0.6681587100028992)]
```

6. In the next two steps, we compute a sentence vector by averaging all the word vectors in the sentence. One of the challenges of this method is representing words that are not present in the model, and here, we simply skip such words. Let's define a function that will take a sentence and a model and return a list of the sentence word vectors. Words that are not present in the model will return KeyError, and in such a case, we catch the error and continue:

```
def get_word_vectors(sentence, model):
    word_vectors = []
    for word in sentence:
        try:
            word_vector = model[word.lower()]
            word_vectors.append(word_vector)
        except KeyError:
            continue
    return word_vectors
```

7. Let's now define a function that will take the word vector list and compute the sentence vector. In order to compute the average, we represent the matrix as a numpy array and use the numpy mean function to get the average vector:

```
def get_sentence_vector(word_vectors):
    matrix = np.array(word_vectors)
    centroid = np.mean(matrix[:,:], axis=0)
    return centroid
```

> **Note**
>
> Averaging the word vectors to get the sentence vector is only one way of approaching this task and is not without its problems. One other alternative is to train a doc2vec model, where sentences, paragraphs, and whole documents can all be units instead of words.

8. We can now test the average word embedding as a vectorizer. Our vectorizer takes in a string input, gets the word vectors for each word, and then returns the sentence vector that we compute in the get_sentence_vector function. We then load the training and test data and create the datasets. We train the logistic regression classifier and test it:

```
vectorize = lambda x: get_sentence_vector(
    get_word_vectors(x, model))
(train_df, test_df) = load_train_test_dataset_pd()
(X_train, X_test, y_train, y_test) = create_train_test_data(
    train_df, test_df, vectorize)
clf = train_classifier(X_train, y_train)
test_classifier(test_df, clf)
```

We can see that the scores are much lower than those in previous sections. There might be several reasons for this; one of them is that the word2vec model is English-only, and the data is multilingual. As an exercise, you can write a script to filter English-only reviews and see whether that improves the score:

	precision	recall	f1-score	support
0	0.54	0.57	0.55	160
1	0.54	0.51	0.53	160
accuracy			0.54	320
macro avg	0.54	0.54	0.54	320
weighted avg	0.54	0.54	0.54	320

There's more...

There are some other fun things that gensim can do with a pretrained model. For example, it can find an outlier word from a list of words and find the word that is most similar to the given word from a list. Let's look at these:

1. Compile a list of words with one that doesn't match, apply the doesnt_match function to the list, and print the results:

    ```
    words = ['banana', 'apple', 'computer', 'strawberry']
    print(model.doesnt_match(words))
    ```

    The result will be as follows:

    ```
    computer
    ```

2. Now, let's find a word that's most similar to another word.

    ```
    word = "cup"
    words = ['glass', 'computer', 'pencil', 'watch']
    print(model.most_similar_to_given(word, words))
    ```

    The result will be as follows:

    ```
    glass
    ```

See also

*   There are many other pretrained models available, including some in other languages; see http://vectors.nlpl.eu/repository/.

    Some pretrained models include part-of-speech information, which can be helpful when disambiguating words. These models concatenate words with their **part of speech** (**POS**) (e.g. cat_NOUN), so keep that in mind when using them.

*   To learn more about the theory behind word2vec, you can start here: https://jalammar.github.io/illustrated-word2vec/.

Training your own embeddings model

We can now train our own word2vec model on a corpus. This model is a neural network that predicts a word when given a sentence with words blanked out. The byproduct of the neural network training is the vector representation for each word in the training vocabulary. For this task, we will continue using the Rotten Tomatoes reviews. The dataset is not very large, so the results are not as good as they could be with a larger collection.

Getting ready

We will use the gensim package for this task. It should be installed as part of the poetry environment.

How to do it...

We will create the dataset and then train the model on the data. We will then test how it performs:

1. Import the necessary packages and functions:

```
import gensim
from gensim.models import Word2Vec
from datasets import load_dataset
from gensim import utils
```

2. Load the training data and check its length:

```
train_dataset = load_dataset("rotten_tomatoes", split="train")
print(len(train_dataset))
```

The result should be as follows:

```
8530
```

3. Create the RottenTomatoesCorpus class. The word2vec training algorithm requires a class with a defined __iter__ function that allows you to iterate through the data, so that is why we need this class:

```
class RottenTomatoesCorpus:

    def __init__(self, sentences):
        self.sentences = sentences

    def __iter__(self):
        for review in self.sentences:
            yield utils.simple_preprocess(
                gensim.parsing.preprocessing.remove_stopwords(
                    review))
```

4. Create an instance of RottenTomatoesCorpus using the loaded training dataset. Since word2vec models are trained on text only (they are self-supervised models), we don't need the review score:

```
sentences = train_dataset["text"]
corpus = RottenTomatoesCorpus(sentences)
```

5. In this step, we initialize the word2vec model, train it, and then save it to disk. The only required argument is the list of words; some of the other important ones are min_count,

size, window, and workers. The min_count parameter refers to the minimum number of times a word has to occur in the training corpus, the default being 5. The size parameter sets the size of the word vector. window restricts the maximum number of words between the predicted and current words in a sentence. workers refers to the number of working threads; the more there are, the quicker the training will proceed. When training the model, the epoch parameter will determine the number of training iterations the model will go through. After initializing the model object, we train it on our corpus for 100 epochs and, finally, save it to disk:

```
model = Word2Vec(sentences=corpus, vector_size=100,
    window=5, min_count=1, workers=4)
model.train(corpus_iterable=corpus,
    total_examples=model.corpus_count, epochs=100)
model.save("../data/rotten_tomato_word2vec.model")
```

6. Get 10 words similar to the word *movie*. The words *sequels* and *film* make sense with this word; the rest are not that related. This is due to the small size of the training corpus. The words you get will be different, since the results are different every time the model is trained:

```
w1 = "movie"
words = model.wv.most_similar(w1, topn=10)
print(words)
```

This is a possible output:

```
[('sequels', 0.38357362151145935), ('film',
0.33577531576156616), ('stuffed', 0.2925359606742859),
('quirkily', 0.28789234161376953), ('convict',
0.2810690104961395), ('worse', 0.2789292335510254), ('churn',
0.27702808380126953), ('hellish', 0.27698105573654175), ('hey',
0.27566075325012207), ('happens', 0.27498629689216614)]
```

There's more...

There are tools to evaluate a word2vec model, although its creation is unsupervised. gensim comes with a file that lists word analogies, such as what *Athens* is to *Greece* being the same as what *Moscow* is to *Russia*. The evaluate_word_analogies function runs the analogies through the model and calculates how many were correct.

Here is how to do this:

1. Use the evaluate_word_analogies function to evaluate our trained model. We need the analogies file, which is available in the book GitHub repository at https://github.com/PacktPublishing/Python-Natural-Language-Processing-Cookbook-Second-Edition/blob/main/data/questions-words.txt

```
(analogy_score, word_list) = model.wv.evaluate_word_analogies(
    '../data/questions-words.txt')
print(analogy_score)
```

The result should be similar to this:

```
0.0015881418740074113
```

2. Let's now evaluate the pretrained model. These commands might take longer to run:

```
pretrained_model = \
    gensim.models.KeyedVectors.load_word2vec_format(
        '../data/GoogleNews-vectors-negative300.bin.gz',
        binary=True)
(analogy_score, word_list) = \
    pretrained_model.evaluate_word_analogies(
        '../data/questions-words.txt')
print(analogy_score)
```

The result should be similar to this:

```
0.7401448525607863
```

3. We use the `evaluate_word_analogies` function differently in the pretrained model and our model case because they are different types. With the pretrained model, we just load the vectors (a `KeyedVectors` class, where each word, represented by a key, is mapped to a vector), and our model is a full `word2vec` model object. We can check the types by using these commands:

```
print(type(pretrained_model))
print(type(model))
```

The result will be as follows:

```
<class 'gensim.models.keyedvectors.KeyedVectors'>
<class 'gensim.models.word2vec.Word2Vec'>
```

The pretrained model was trained on a much larger corpus and, predictably, performs better. You can also construct your own evaluation file with analogies that your data requires.

> **Note**
>
> Make sure your evaluation is based on the type of data that you are going to use in your application; otherwise, you risk having misleading evaluation results.

See also

There is an additional way of evaluating model performance, by comparing the similarity between word pairs assigned by a model to the human-assigned judgments. You can do this by using the `evaluate_word_pairs` function. See more at https://radimrehurek.com/gensim/models/keyedvectors.html#gensim.models.keyedvectors.KeyedVectors.evaluate_word_pairs.

Using BERT and OpenAI embeddings instead of word embeddings

Instead of word embeddings, we can use **Bidirectional Encoder Representations from Transformer** (**BERT**) embeddings. A BERT model, like word embeddings, is a pretrained model and gives a vector representation, but it takes context into account and can represent a whole sentence instead of individual words.

Getting ready

For this recipe, we can use the Hugging Face `sentence_transformers` package to represent sentences as vectors. We need `PyTorch`, which is installed as part of the `poetry` environment.

To get the vectors, we will use the `all-MiniLM-L6-v2` model for this recipe.

We can also use the embeddings from OpenAI that come from their **large language models** (**LLMs**).

To use the OpenAI embeddings, you will need to create an account and get an API key from OpenAI. You can create an account at `https://platform.openai.com/signup`.

The notebook is located at `https://github.com/PacktPublishing/Python-Natural-Language-Processing-Cookbook-Second-Edition/blob/main/Chapter03/3.6_train_own_word2vec.ipynb`.

How to do it...

Hugging Face code makes using BERT very easy. The first time the code runs, it will download the necessary model, which might take some time. After the download, it's just a matter of encoding the sentences using the model. We will test the simple classifier with these embeddings:

1. Run the simple classifier notebook to import its functions:

    ```
    %run -i "../util/util_simple_classifier.ipynb"
    ```

2. Import the `SentenceTransformer` class:

    ```
    from sentence_transformers import SentenceTransformer
    ```

3. Load the sentence transformer model, retrieve the embedding of the sentence *I love jazz*, and print it out.

    ```
    model = SentenceTransformer('all-MiniLM-L6-v2')
    embedding = model.encode(["I love jazz"])
    print(embedding)
    ```

As we can see, it is a vector similar to the word embeddings vector from the previous recipe:

```
[[ 2.94217980e-03 -7.93536603e-02 -2.82228496e-02 -5.13779782e-
02
  -6.44981042e-02  9.83557850e-02  1.09671958e-01 -3.26390602e-
02
   4.96566631e-02  2.56580133e-02 -1.08482063e-01  1.88441798e-
02
   2.70963665e-02 -3.80690470e-02  2.42502335e-02 -3.65605950e-
03
   1.29364491e-01  4.32255343e-02 -6.64561391e-02 -6.93060979e-
02
  -1.39410645e-01  4.36719768e-02 -7.85463024e-03  1.68625098e-
02
  -1.01160072e-02  1.07926019e-02 -1.05814040e-02  2.57284809e-
02
  -1.51516097e-02 -4.53920700e-02  7.12087378e-03  1.17573030e-
01… ]]
```

4. Now, we can test our classifier using the BERT embeddings. First, let's define a function that will return a sentence vector. This function takes the input text and a model. It then uses the model to encode the text and returns the resulting embedding. We need to pass in the text inside of a list to the `encode` method, as it expects an iterable. Similarly, we return the first element of the result, since it returns a list of embeddings.

```
def get_sentence_vector(text, model):
    sentence_embeddings = model.encode([text])
    return sentence_embeddings[0]
```

5. Now, we define the `vectorize` function, create the training and test data using the `load_train_test_dataset_pd` function we created in the *Creating a simple classifier* recipe, train the classifier, and test it. We will time the dataset creation step, hence the inclusion of the `time` package commands. We see that it takes about 11 seconds to vectorize the whole dataset (about 85,000 entries). We then train the model and test it:

```
import time
vectorize = lambda x: get_sentence_vector(x, model)
(train_df, test_df) = load_train_test_dataset_pd()
start = time.time()
(X_train, X_test, y_train, y_test) = create_train_test_data(
    train_df, test_df, vectorize)
print(f"BERT embeddings: {time.time() - start} s")
clf = train_classifier(X_train, y_train)
test_classifier(test_df, clf)
```

The result is our best one yet:

```
BERT embeddings: 11.410213232040405 s
              precision    recall  f1-score   support

           0      0.77      0.79      0.78       160
           1      0.79      0.76      0.77       160

    accuracy                          0.78       320
   macro avg      0.78      0.78      0.78       320
weighted avg      0.78      0.78      0.78       320
```

There's more...

We can now use the OpenAI embeddings to see how they perform:

1. Import the openai package and assign the API key:

    ```
    import openai
    openai.api_key = OPEN_AI_KEY
    ```

2. Assign the model we are going to use, the sentence and create the embedding. The model that we will use is specifically an embeddings model, so it returns an embeddings vector for a text input:

    ```
    model = "text-embedding-ada-002"
    text = "I love jazz"
    response = openai.Embedding.create(
        input=text,
        model=model
    )
    embeddings = response['data'][0]['embedding']
    print(embeddings)
    ```

    The partial result will be as follows:

    ```
    [-0.028350897133350372, -0.011136125773191452,
    -0.0021299426443874836, -0.014453398995101452,
    -0.012048527598381042, 0.018223850056529045,
    -0.010247894562780857, -0.01806674897670746,
    -0.014308380894362926, 0.0007220656843855977,
    -9.998268797062337e-05, 0.010078707709908485,…]
    ```

3. Let's now test our classifier using the OpenAI embeddings. This is the function that will return a sentence vector:

    ```
    def get_sentence_vector(text, model):
        text = "I love jazz"
        response = openai.Embedding.create(
    ```

```
        input=text,
        model=model
    )
    embeddings = response['data'][0]['embedding']
    return embeddings
```

4. Now, define the `vectorize` function, create the training and test data, train the classifier, and test it. We will time the vectorizing step:

```
import time
vectorize = lambda x: get_sentence_vector(x, model)
(train_df, test_df) = load_train_test_dataset_pd()
start = time.time()
(X_train, X_test, y_train, y_test) = create_train_test_data(
    train_df, test_df, vectorize)
print(f"OpenAI embeddings: {time.time() - start} s")
clf = train_classifier(X_train, y_train)
test_classifier(test_df, clf)
```

The result will be as follows:

```
OpenAI embeddings: 704.3250799179077 s
              precision    recall  f1-score   support

           0       0.49      0.82      0.62       160
           1       0.47      0.16      0.23       160

    accuracy                           0.49       320
   macro avg       0.48      0.49      0.43       320
weighted avg       0.48      0.49      0.43       320
```

Note that the result is quite poor in terms of the score, and it takes more than 10 minutes to process the whole dataset. Here, we only use the LLM embeddings and then train a logistic regression classifier on those embeddings. This is different from using the LLM itself to do the classification.

See also

For more pretrained models, see https://www.sbert.net/docs/pretrained_models.html.

Retrieval augmented generation (RAG)

In this recipe, we will see vector embeddings in action. RAG is a popular method of working with LLMs. Since these models are pretrained on widely available internet data, they do not have access to our personal data, and we cannot use the model as it is to ask questions about it. A way to overcome this is to use vector embeddings to represent our data. Then, we can compute cosine similarity between our data and the question and include the most similar piece of our data, together with the question – hence the name "retrieval augmented generation," since we first retrieve relevant data by using cosine similarity and then generate text using the LLM.

Getting ready

We will use an IMDB dataset from **Kaggle**, which can be downloaded from `https://www.kaggle.com/PromptCloudHQ/imdb-data` and is also included in the book GitHub repo at `https://github.com/PacktPublishing/Python-Natural-Language-Processing-Cookbook-Second-Edition/blob/main/data/IMDB-Movie-Data.csv`. Download the dataset and unzip the CSV file.

We will also use the OpenAI embeddings, as well as the `llama_index` package, which is included within the `poetry` environment.

The notebook is located at `https://github.com/PacktPublishing/Python-Natural-Language-Processing-Cookbook-Second-Edition/blob/main/Chapter03/3.9_vector_search.ipynb`.

How to do it...

We will load the IMDB dataset and then create a vector store, using its first 10 entries. We will then use the `llama_index` package to query the vector store:

1. Run the file `utilities` notebook:

    ```
    %run -i "../util/file_utils.ipynb"
    ```

2. Import the necessary classes and packages:

    ```
    import csv
    import openai
    from llama_index import VectorStoreIndex
    from llama_index import Document
    openai.api_key = OPEN_AI_KEY
    ```

3. Read in the CSV data. We will skip the first header row of the data:

```
with open('../data/IMDB-Movie-Data.csv') as f:
    reader = csv.reader(f)
    data = list(reader)
    movies = data[1:]
```

4. In this step, we use the first 10 rows of the data we just read in to first create a list of Document objects, and then a VectorStoreIndex object with these Document objects. An index is an object used for search, where each record contains certain information. A vector store index stores metadata as well as the vector representation of each record. For each movie, we assign the description as the text that will be embedded and the rest as metadata. We print out document objects and can see that a unique ID has been assigned to each:

```
documents = []
for movie in movies[0:10]:
    doc_id = movie[0]
    title = movie[1]
    genres = movie[2].split(",")
    description = movie[3]
    director = movie[4]
    actors = movie[5].split(",")
    year = movie[6]
    duration = movie[7]
    rating = movie[8]
    revenue = movie[10]
    document = Document(
        text=description,
        metadata={
            "title": title,
            "genres": genres,
            "director": director,
            "actors": actors,
            "year": year,
            "duration": duration,
            "rating": rating,
            "revenue": revenue
        }
    )
    print(document)
    documents.append(document)
index = VectorStoreIndex.from_documents(documents)
```

The partial output will look like this:

```
id_='6e1ef633-f10b-44e3-9b77-f5f7b08dcedd' embedding=None
metadata={'title': 'Guardians of the Galaxy', 'genres':
['Action', 'Adventure', 'Sci-Fi'], 'director': 'James
Gunn', 'actors': ['Chris Pratt', ' Vin Diesel', ' Bradley
Cooper', ' Zoe Saldana'], 'year': '2014', 'duration': '121',
'rating': '8.1', 'revenue': '333.13'} excluded_embed_metadata_
keys=[] excluded_llm_metadata_keys=[] relationships={}
hash='e18bdce3a36c69d8c1e55a7eb56f05162c68c97151cbaf40

91814ae3df42dfe8' text='A group of intergalactic criminals
are forced to work together to stop a fanatical warrior from
taking control of the universe.' start_char_idx=None end_char_
idx=None text_template='{metadata_str}\n\n{content}' metadata_
template='{key}: {value}' metadata_seperator='\n'
```

5. Create the query engine from the index we just created. The query engine will allow us to send in questions about the documents loaded in the index:

```
query_engine = index.as_query_engine()
```

6. Use the engine to answer a question:

```
response = query_engine.query("""Which movies talk about
something gigantic?""")
```

```
print(response.response)
```

The answer seems to make sense grammatically, and arguably the Great Wall of China is gigantic. However, it is not clear what is gigantic in the movie Prometheus. So here we have a partially correct answer. The Great Wall and Prometheus both talk about something gigantic. In The Great Wall, the protagonists become embroiled in the defense of the Great Wall of China against a horde of monstrous creatures. In Prometheus, the protagonists find a structure on a distant moon.

4

Classifying Texts

In this chapter, we will be classifying texts using different methods. Classifying texts is a classic NLP problem. This NLP task involves assigning a value to a text, for example, a topic (such as sport or business) or a sentiment, such as negative or positive, and any such task needs evaluation.

After reading this chapter, you will be able to preprocess and classify texts using keywords, unsupervised clustering, and two supervised algorithms: **support vector machines** (**SVMs**) and a **convolutional neural network** (**CNN**) model trained within the spaCy framework. We will also use GPT-3.5 to classify texts.

For theoretical background on some of the concepts discussed in this section, please refer to *Building Machine Learning Systems with Python* by Coelho et al. That book will explain the basics of building a machine learning project, such as training and test sets, as well as metrics used to evaluate such projects, including precision, recall, F1, and accuracy.

Here is a list of the recipes in this chapter:

- Getting the dataset and evaluation ready
- Performing rule-based text classification using keywords
- Clustering sentences using K-Means – unsupervised text classification
- Using SVMs for supervised text classification
- Training a spaCy model for supervised text classification
- Classifying texts using OpenAI models

Technical requirements

The code for this chapter can be found in the `Chapter04` folder in the GitHub repository of the book (`https://github.com/PacktPublishing/Python-Natural-Language-Processing-Cookbook-Second-Edition`). As always, we will use the `poetry` environment to install the necessary packages. You can also install the required packages using the provided `requirements.txt` file. We will use the Hugging Face `datasets` package to get datasets that we will use throughout the chapter.

Getting the dataset and evaluation ready

In this recipe, we will load a dataset, prepare it for processing, and create an evaluation baseline. This recipe builds on some of the recipes from *Chapter 3*, where we used different tools to represent text in a computer-readable form.

Getting ready

For this recipe, we will use the Rotten Tomatoes reviews dataset, available through Hugging Face. This dataset consists of user movie reviews that can be classified into positive and negative. We will prepare the dataset for machine learning classification. The preparation process in this case will involve loading the reviews, filtering out non-English language ones, tokenizing the text into words, and removing stopwords. Before the machine learning algorithm can run, the text reviews need to be transformed into vectors. This transformation process is described in detail in *Chapter 3*.

The notebook is located at `https://github.com/PacktPublishing/Python-Natural-Language-Processing-Cookbook-Second-Edition/blob/main/Chapter04/4.1_data_preparation.ipynb`.

How to do it...

We will classify whether the input review is of negative or positive sentiment. We will first filter out non-English text, then tokenize it into words and remove stopwords and punctuation. Finally, we will look at the class distribution and review the most common words in each class.

Here are the steps:

1. Run the simple classifier file:

   ```
   %run -i "../util/util_simple_classifier.ipynb"
   ```

2. Import necessary classes. We import the `detect` function from `langdetect`, which will help us determine the language of the review. We also import the `word_tokenize` function, which we will use to split the reviews into words. The `FreqDist` class from NLTK will help us see the most frequent positive and negative words in the reviews. We will use the `stopwords` list, also from NLTK, to filter the stopwords from the text. Finally, the `punctuation` string from the `string` package will help us to filter punctuation:

   ```
   from langdetect import detect
   from nltk import word_tokenize
   from nltk.probability import FreqDist
   from nltk.corpus import stopwords
   from string import punctuation
   ```

3. Load the training and test datasets using the function from the simple classifier file and print the two dataframes. We see that the data contains a `text` column and a `label` column, where the text column is lowercase:

```
(train_df, test_df) = load_train_test_dataset_pd("train",
    "test")
print(train_df)
print(test_df)
```

The output should look similar to this:

```
                                                    text  label
0       the rock is destined to be the 21st century's ...      1
1       the gorgeously elaborate continuation of " the...      1
...                                                   ...    ...
8525    any enjoyment will be hinge from a personal th...      0
8526    if legendary shlockmeister ed wood had ever ma...      0
[8530 rows x 2 columns]
                                                    text  label
0       lovingly photographed in the manner of a golde...      1
1                      consistently clever and suspenseful .      1
...                                                   ...    ...
1061    a terrible movie that some people will neverth...      0
1062    there are many definitions of 'time waster' bu...      0
[1066 rows x 2 columns]
```

4. Now we create a new column called `lang` in the dataframes that will contain the language of the review. We use the `detect` function to populate this column via the `apply` method. We then filter the dataframe to only contain English-language reviews. The final row counts of the training dataframe before and after the filtering show us that 178 rows were non-English. This step may take a minute to run:

```
train_df["lang"] = train_df["text"].apply(detect)
train_df = train_df[train_df['lang'] == 'en']
print(train_df)
```

Now the output should look something like this:

```
                                                    text  label
lang
0       the rock is destined to be the 21st century's
...        1    en
1       the gorgeously elaborate continuation of "
the...        1    en
...                                                   ...
      ...    ...
8528    interminably bleak , to say nothing of boring
```

```
.            0    en
8529  things really get weird , though not
particula...        0    en


[8364 rows x 3 columns]
```

5.   Now we will do the same for the test dataframe:

```
test_df["lang"] = test_df["text"].apply(detect)
test_df = test_df[test_df['lang'] == 'en']
```

6.   Now we will tokenize the text into words. If you get an error saying that the english.pickle tokenizer was not found, run the line nltk.download('punkt') before running the rest of the code. This code is also contained in the lang_utils notebook (https://github.com/PacktPublishing/Python-Natural-Language-Processing-Cookbook-Second-Edition/blob/main/util/lang_utils.ipynb):

```
train_df["tokenized_text"] = train_df["text"].apply(
    word_tokenize)
print(train_df)
test_df["tokenized_text"] = test_df["text"].apply(word_tokenize)
print(test_df)
```

The result will be similar to this:

```
                                                  text    label
lang  \
0        the rock is destined to be the 21st century's ...
         1    en
1        the gorgeously elaborate continuation of " the...
1    en
...                                                  ...      ...
   ...
8528     interminably bleak , to say nothing of boring .
0    en
8529  things really get weird , though not particula...
0    en


                                            tokenized_text
0        [the, rock, is, destined, to, be, the, 21st, c...
1        [the, gorgeously, elaborate, continuation, of,...
...                                                  ...
8528  [interminably, bleak, ,, to, say, nothing, of,...
8529  [things, really, get, weird, ,, though, not, p...


[8352 rows x 4 columns]
```

7. In this step, we will remove stopwords and punctuation. First, we load the stopwords using the NLTK package. We then add 's and `` to the list of stopwords. You can add other words that you think are also stopwords. We then define a function that will take a list of words as input and filter it, returning a new list that doesn't contain stopwords or punctuation. Finally, we apply this function to the training and test data. From the printout, we can see that stopwords and punctuation were removed:

```
stop_words = list(stopwords.words('english'))
stop_words.append("``")
stop_words.append("'s")
def remove_stopwords_and_punct(x):
    new_list = [w for w in x if w not in stop_words and w not in
punctuation]
    return new_list
train_df["tokenized_text"] = train_df["tokenized_text"].apply(
    remove_stopwords_and_punct)
print(train_df)
test_df["tokenized_text"] = test_df["tokenized_text"].apply(
    remove_stopwords_and_punct)
print(test_df)
```

The result will look similar to this:

```
                                              text  label
lang  \
0      the rock is destined to be the 21st century's ...
       1    en
1      the gorgeously elaborate continuation of " the...
       1    en
...                                              ...
      ...   ...
8528    interminably bleak , to say nothing of boring .
       0    en
8529  things really get weird , though not particula...
       0    en

                                    tokenized_text
0      [rock, destined, 21st, century, new, conan, go...
1      [gorgeously, elaborate, continuation, lord, ri...
...                                              ...
8528         [interminably, bleak, say, nothing, boring]
8529  [things, really, get, weird, though, particula...

[8352 rows x 4 columns]
```

8. Now we will check the class balance over both datasets. It is important that the number of items in each class is approximately the same, since if one class dominates, the model can just learn to always assign this dominating class without being wrong much of the time:

```
print(train_df.groupby('label').count())
print(test_df.groupby('label').count())
```

We see that there are slightly, but not significantly, more negative reviews in the training data than positive, and the numbers are nearly equal in test data.

```
        text   lang   tokenized_text
label
0       4185   4185             4185
1       4167   4167             4167
        text   lang   tokenized_text
label
0        523    523              523
1        522    522              522
```

9. Let's now save the cleaned data to disk:

```
train_df.to_json("../data/rotten_tomatoes_train.json")
test_df.to_json("../data/rotten_tomatoes_test.json")
```

10. In this step, we define a function that will take a list of words and the number of words as input and return a `FreqDist` object. It will also print out the top *n* most frequent words, where *n* is passed into the function and is 200 by default:

```
def get_stats(word_list, num_words=200):
    freq_dist = FreqDist(word_list)
    print(freq_dist.most_common(num_words))
    return freq_dist
```

11. Now let's use the preceding function and show the most common words in positive and negative reviews to see whether there are significant vocabulary differences between the two classes. We create two lists of words, one for positive and one for negative reviews. We first filter the dataframe by label and then use the `sum` function to get the words from all the reviews:

```
positive_train_words = train_df[
    train_df["label"] == 1].tokenized_text.sum()
negative_train_words = train_df[
    train_df["label"] == 0].tokenized_text.sum()
positive_fd = get_stats(positive_train_words)
negative_fd = get_stats(negative_train_words)
```

In the output, we see that the words `film` and `movie` and some other words also act as stopwords in this case, as they are the most common words in both sets. We can add them to the stopwords list in step 7 and redo the cleaning:

```
[('film', 683), ('movie', 429), ("n't", 286), ('one', 280),
('--', 271), ('like', 209), ('story', 194), ('comedy', 160),
('good', 150), ('even', 144), ('funny', 137), ('way', 135),
('time', 127), ('best', 126), ('characters', 125), ('make',
124), ('life', 124), ('much', 122), ('us', 122), ('love', 118),
...]
[('movie', 641), ('film', 557), ("n't", 450), ('like', 354),
('one', 293), ('--', 264), ('story', 189), ('much', 177),
('bad', 173), ('even', 160), ('time', 146), ('good', 143),
('characters', 138), ('little', 137), ('would', 130), ('never',
122), ('comedy', 121), ('enough', 107), ('really', 105),
('nothing', 103), ('way', 102), ('make', 101), ...]
```

Performing rule-based text classification using keywords

In this recipe, we will use the vocabulary of the text to classify the Rotten Tomatoes reviews. We will create a simple classifier that will have a vectorizer for each class. That vectorizer will include the words characteristic to that class. The classification will simply be vectorizing the text using each of the vectorizers and then using the class that has more words.

Getting ready

We will use the `CountVectorizer` class and the `classification_report` function from `sklearn`, as well as the `word_tokenize` method from NLTK. All of these are included in the `poetry` environment.

The notebook is located at `https://github.com/PacktPublishing/Python-Natural-Language-Processing-Cookbook-Second-Edition/blob/main/Chapter04/4.2_rule_based.ipynb`.

How to do it...

In this recipe, we will create a separate vectorizer for each class. We will then use those vectorizers to count the number of each class word in each review to classify it:

1. Run the simple classifier file:

    ```
    %run -i "../util/util_simple_classifier.ipynb"
    ```

2. Do the necessary imports:

    ```
    from nltk import word_tokenize
    from sklearn.feature_extraction.text import CountVectorizer
    from sklearn.metrics import classification_report
    ```

3. Load the cleaned data from disk. If you receive a `FileNotFoundError` error at this step, you need to run the previous recipe, *Getting the dataset and evaluation ready*, first, since those files were created there after cleaning the data:

```
train_df = pd.read_json("../data/rotten_tomatoes_train.json")
test_df = pd.read_json("../data/rotten_tomatoes_test.json")
```

4. Here we create a list of words unique to each class. We first concatenate all the words from the `text` column, filtering on the relevant `label` value (0 for negative reviews and 1 for positive ones). We then get the words that appear in both of those lists in the `word_intersection` variable. Finally, we create filtered word lists, one for each class, that do not contain words that appear in both classes. Basically, we delete all the words that appear in both positive and negative reviews from the respective lists:

```
positive_train_words = train_df[train_df["label"]
    == 1].text.sum()
negative_train_words = train_df[train_df["label"]
    == 0].text.sum()
word_intersection = set(positive_train_words) \
    & set(negative_train_words)
positive_filtered = list(set(positive_train_words)
    - word_intersection)
negative_filtered = list(set(negative_train_words)
    - word_intersection)
```

5. Next, we define a function to create vectorizers, one for each class. The input to this function is a list of lists, where each one is a list of words that only appear in that class; we created these in the previous step. For each of the word lists, we create a `CountVectorizer` object that takes the word list as the `vocabulary` parameter. Providing this ensures that we only count those words for the purpose of classification:

```
def create_vectorizers(word_lists):
    vectorizers = []
    for word_list in word_lists:
        vectorizer = CountVectorizer(vocabulary=word_list)
        vectorizers.append(vectorizer)
    return vectorizers
```

6. Create the vectorizers using the preceding function:

```
vectorizers = create_vectorizers([negative_filtered,
    positive_filtered])
```

7. In this step, we create a `vectorize` function that takes in a list of words and a list of vectorizers. We first create a string from the word list, as the vectorizer expects a string. For each vectorizer in the list, we apply it to the text and then sum the total count of words in that vectorizer. Finally, we append that sum to a list of scores. This will count words in the input per class. We return this score list at the end of the function:

```
def vectorize(text_list, vectorizers):
    text = " ".join(text_list)
    scores = []
    for vectorizer in vectorizers:
        output = vectorizer.transform([text])
        output_sum = sum(output.todense().tolist()[0])
        scores.append(output_sum)
    return scores
```

8. In this step, we define the `classify` function, which takes a list of scores returned by the `vectorize` function. This function simply selects the maximum score from the list and returns the index of that score corresponding to the class label:

```
def classify(score_list):
    return max(enumerate(score_list),key=lambda x: x[1])[0]
```

9. Here, we apply the preceding functions to the training data. We first vectorize the text and then classify it. We create a new column for the result called `prediction`:

```
train_df["prediction"] = train_df["text"].apply(
    lambda x: classify(vectorize(x, vectorizers)))
print(train_df)
```

The output will look similar to this:

```
                                                   text   label
lang   \
0      [rock, destined, 21st, century, new, conan, go...
1    en
1      [gorgeously, elaborate, continuation, lord, ri...
1    en
...                                                 ...     ...
  ...
8528         [interminably, bleak, say, nothing, boring]
0    en
8529  [things, really, get, weird, though, particula...
0    en

       prediction
0               1
```

```
1                    1
...                  ...
8528                 0
8529                 0

[8364 rows x 4 columns]
```

10. Now we measure the performance of the rule-based classifier by printing the classification report. We input the assigned label and the prediction columns. The result is an overall accuracy score of 87%:

```
print(classification_report(train_df['label'],
    train_df['prediction']))
```

This results in the following:

	precision	recall	f1-score	support
0	0.79	0.99	0.88	4194
1	0.99	0.74	0.85	4170
accuracy			0.87	8364
macro avg	0.89	0.87	0.86	8364
weighted avg	0.89	0.87	0.86	8364

11. Here we do the same for the test data, and we see a significant reduction in accuracy, down to 62%. This is because the vocabulary lists that we use to create the vectorizers only come from the training data and are not exhaustive. They will lead to errors in unseen data:

```
test_df["prediction"] = test_df["text"].apply(
    lambda x: classify(vectorize(x, vectorizers)))
print(classification_report(test_df['label'],
    test_df['prediction']))
```

The result will be as follows:

	precision	recall	f1-score	support
0	0.59	0.81	0.68	523
1	0.70	0.43	0.53	524
accuracy			0.62	1047
macro avg	0.64	0.62	0.61	1047
weighted avg	0.64	0.62	0.61	1047

Clustering sentences using K-Means – unsupervised text classification

In this recipe, we will use the BBC news dataset. The dataset contains news pieces sorted by five topics: politics, tech, business, sport, and entertainment. We will apply the unsupervised K-Means algorithm to sort the data into unlabeled classes.

After you read this recipe, you will be able to create your own unsupervised clustering model that will sort data into several classes. You can then later apply it to any text data without having to first label it.

Getting ready

We will use the KMeans algorithm to create our unsupervised model. It is part of the sklearn package and is included in the poetry environment.

The BBC news dataset as we use it here was uploaded by a Hugging Face user, and the link and the dataset might change in time. To avoid any potential issues, you can use the BBC dataset uploaded to the book's GitHub repository by loading it from the CSV file provided in the data directory.

The notebook is located at https://github.com/PacktPublishing/Python-Natural-Language-Processing-Cookbook-Second-Edition/blob/main/Chapter04/4.3_unsupervised_classification.ipynb.

How to do it...

In this recipe, we will preprocess the data, vectorize it, and then cluster it using K-Means. Since there are usually no right answers for unsupervised modeling, evaluating the models is more difficult, but we will be able to look at some statistics, as well as the most common words in all the clusters.

Your steps should be formatted like so:

1. Run the simple classification file:

```
%run -i "../util/util_simple_classifier.ipynb"
%run -i "../util/lang_utils.ipynb"
```

2. Import the necessary functions and packages:

```
from nltk import word_tokenize
from sklearn.cluster import KMeans
from nltk.probability import FreqDist
from sklearn.feature_extraction.text import TfidfVectorizer
from sklearn.model_selection import StratifiedShuffleSplit
```

3. We will load the BBC dataset. We use the `load_dataset` function from Hugging Face's `datasets` package. This function was imported in the simple classifier file we ran in step 1. In the Hugging Face repository, datasets are usually split into training and testing. We will load both, although in unsupervised learning, the test set is usually not used:

```
train_dataset = load_dataset("SetFit/bbc-news", split="train")
test_dataset = load_dataset("SetFit/bbc-news", split="test")
train_df = train_dataset.to_pandas()
test_df = test_dataset.to_pandas()
print(train_df)
print(test_df)
```

The result will look similar to this:

```
                                                   text   label
      label_text
0     wales want rugby league training wales could f...      2
         sport
1     china aviation seeks rescue deal scandal-hit j...      1
         business
...                                                ...    ...

         ...
1223  why few targets are better than many the econo...     1
         business
1224  boothroyd calls for lords speaker betty boothr...     4
         politics

[1225 rows x 3 columns]
                                                   text   label
      label_text
0     carry on star patsy rowlands dies actress pats...      3
   entertainment
1     sydney to host north v south game sydney will ...      2
         sport
..                                                 ...    ...

         ...
998   stormy year for property insurers a string of ...     1
         business
999   what the election should really be about  a ge...    4
         politics

[1000 rows x 3 columns]
```

4. Now we will check the distribution of items per class for both training and test data. Class balance is important in classification, as a disproportionally larger class will influence the final classifier:

```
print(train_df.groupby('label_text').count())
print(test_df.groupby('label_text').count())
```

We see that the classes are pretty evenly split, but there are more examples in the business and sport categories:

	text	label
label_text		
business	286	286
entertainment	210	210
politics	242	242
sport	275	275
tech	212	212
	text	label
label_text		
business	224	224
entertainment	176	176
politics	175	175
sport	236	236
tech	189	189

5. Since there is almost as much data in the test set as in the training set, we will combine the data and create a better train/test split. We first concatenate the two dataframes. We then create a StratifiedShuffleSplit that will create a train/test split and will do it while preserving the class balance. We specify that we only need one split (n_splits) and that the test data needs to be 20% of the whole dataset (test_size). The sss object's split method returns a generator that contains the indices for the split. We can then use these indices to get new training and test dataframes. To do that, we filter on the relevant indices and then make a copy of the resulting dataframe slice. If we didn't make a copy, then we would be working on the original dataframe. We then print out the class counts for both dataframes and see that there is more training and less testing data:

```
combined_df = pd.concat([train_df, test_df],
    ignore_index=True, sort=False)
print(combined_df)
sss = StratifiedShuffleSplit(n_splits=1,
    test_size=0.2, random_state=0)
train_index, test_index = next(
    sss.split(combined_df["text"], combined_df["label"]))
train_df = combined_df[combined_df.index.isin(
    train_index)].copy()
test_df = combined_df[combined_df.index.isin(test_index)].copy()
print(train_df.groupby('label_text').count())
print(test_df.groupby('label_text').count())
```

The result should look like this:

label_text	text	label	text_tokenized	text_clean	cluster
business	408	408	408	408	330
entertainment	309	309	309	309	253
politics	333	333	333	333	263
sport	409	409	409	409	327
tech	321	321	321	321	262
label_text	text	label	text_tokenized	text_clean	cluster
business	102	102	102	102	78
entertainment	77	77	77	77	56
politics	84	84	84	84	70
sport	102	102	102	102	82
tech	80	80	80	80	59

6. We will now preprocess the data: tokenize it and remove stopwords and punctuation. The functions to do this (`tokenize`, `remove_stopword_punct`) are imported in the `language_utils` file we ran in step 1. If you get an error that the `english.pickle` tokenizer was not found, run the line `nltk.download('punkt')` before running the rest of the code. This code is also contained in the `lang_utils notebook`:

```
train_df = tokenize(train_df, "text")
train_df = remove_stopword_punct(train_df, "text_tokenized")
test_df = tokenize(test_df, "text")
test_df = remove_stopword_punct(test_df, "text_tokenized")
```

7. In this step, we create the vectorizer. To do that, we get all the words from the training news articles. First, we save the clean text in a separate column, `text_clean`, and then we save the two dataframes to disk. Then we create a TF-IDF vectorizer that will count unigrams, bigrams, and trigrams (the `ngram_range` parameter). We then fit the vectorizer on the training data only. The reason we fit it only on the training data is that if we fit it on both training and test data, it would lead to data leakage and we would get better test scores than actual performance on unseen data:

```
train_df["text_clean"] = train_df["text_tokenized"].apply(
    lambda x: " ".join(list(x)))
test_df["text_clean"] = test_df["text_tokenized"].apply(
    lambda x: " ".join(list(x)))
train_df.to_json("../data/bbc_train.json")
test_df.to_json("../data/bbc_test.json")
vec = TfidfVectorizer(ngram_range=(1,3))
matrix = vec.fit_transform(train_df["text_clean"])
```

8. Now we can create the `KMeans` classifier for five clusters and then fit it on the matrix produced using the vectorizer from the preceding code. We specify the number of clusters using the `n_clusters` parameter. We also specify that the number of times the algorithm should run is 10 using the `n_init` parameter. For higher-dimensional problems, it is recommended to do several runs. After initializing the classifier, we fit it on the matrix we created using the vectorizer in step 7. This will create the clustering of the training data:

> **Note**
>
> In real-life projects, you will not know the number of clusters in advance, as we do here. You will need to use the elbow method or other methods to estimate the optimal number of classes.

```
km = KMeans(n_clusters=5, n_init=10)
km.fit(matrix)
```

9. The `get_most_frequent_words` function will return a list of the most frequent words in a list. The most frequent words list will provide us with a clue as to which topic the text is about. We will use this function to print out the most frequent words in a cluster to understand which topic they refer to. The function takes in input text, tokenizes it, and then creates a `FreqDist` object. We get the top word frequency tuples by using its `most_common` function and finally get only the word without the frequencies and return this as a list:

```
def get_most_frequent_words(text, num_words):
    word_list = word_tokenize(text)
    freq_dist = FreqDist(word_list)
    top_words = freq_dist.most_common(num_words)
    top_words = [word[0] for word in top_words]
    return top_words
```

10. In this step, we define another function, `print_most_common_words_by_cluster`, which uses the `get_most_frequent_words` function we defined in the previous step. We take the dataframe, the `KMeans` model, and the number of clusters as input parameters. We then get the list of the clusters assigned to each data point and then create a column in the dataframe that specifies the assigned cluster. For each cluster, we then filter the dataframe to get the text just for that cluster. We use this text to then pass it into the `get_most_frequent_words` function to get the list of the most frequent words in that cluster. We print the cluster number and the list and return the input dataframe with the added cluster number column:

```
def print_most_common_words_by_cluster(input_df, km,
    num_clusters):
    clusters = km.labels_.tolist()
    input_df["cluster"] = clusters
    for cluster in range(0, num_clusters):
        this_cluster_text = input_df[
            input_df['cluster'] == cluster]
```

```
        all_text = " ".join(
            this_cluster_text['text_clean'].astype(str))
        top_200 = get_most_frequent_words(all_text, 200)
        print(cluster)
        print(top_200)
    return input_df
```

11. Here, we use the function we defined in the previous step on the training dataframe. We also pass in the fitted KMeans model and the number of clusters, 5. The printout gives us an idea of which cluster is which topic. The cluster numbers might vary, but the cluster that has labour, party, election as the most frequent words is the politics cluster; the cluster with the words music, award, and show is the entertainment cluster; the cluster with the words game, England, win, play, and cup is the sport cluster; the cluster with the words sales and growth is the business cluster; and the cluster with the words software, net, and search is the tech cluster. We also note that the words said and Mr are clearly stopwords, as they appear in most clusters close to the top:

```
print_most_common_words_by_cluster(train_df, km, 5)
```

The results will vary each time you run the training, but they might look like this (output truncated):

```
0
['mr', 'said', 'would', 'labour', 'party', 'election', 'blair',
'government', ...]
1
['film', 'said', 'best', 'also', 'year', 'one', 'us', 'awards',
'music', 'new', 'number', 'award', 'show', ...]
2
['said', 'game', 'england', 'first', 'win', 'world', 'last',
'one', 'two', 'would', 'time', 'play', 'back', 'cup', 'players',
...]
3
['said', 'mr', 'us', 'year', 'people', 'also', 'would', 'new',
'one', 'could', 'uk', 'sales', 'firm', 'growth', ...]
4
['said', 'people', 'software', 'would', 'users', 'mr', 'could',
'new', 'microsoft', 'security', 'net', 'search', 'also', ...]
```

12. In this step, we use the fitted model to predict the cluster for a test example. We use the text in row 1 of the test dataframe. It is a politics example. We use the vectorizer to turn the text into a vector and then use the K-Means model to predict the cluster. The prediction is cluster 0, which in this case is correct:

```
test_example = test_df.iloc[1, test_df.columns.get_loc('text')]
print(test_example)
vectorized = vec.transform([test_example])
prediction = km.predict(vectorized)
print(prediction)
```

The result might look like this:

```
lib dems  new election pr chief the lib dems have appointed
a senior figure from bt to be the party s new communications
chief for their next general election effort.  sandy walkington
will now work with senior figures such as matthew taylor on
completing the party manifesto. party chief executive lord
rennard said the appointment was a  significant strengthening
of the lib dem team . mr walkington said he wanted the party
to be ready for any  mischief  rivals or the media tried to
throw at it.   my role will be to ensure this new public profile
is effectively communicated at all levels   he said.  i also
know the party will be put under scrutiny in the media and
from the other parties as never before - and we will need to
show ourselves ready and prepared to counter the mischief and
misrepresentation that all too often comes from the party s
opponents.  the party is already demonstrating on every issue
that it is the effective opposition.  mr walkington s new job
title is director of general election communications.
[0]
```

13. Finally, we save the model using the `joblib` package's `dump` function and then load it again using the `load` function. We check the prediction of the loaded model, and it is the same as the prediction of the model in memory. This step will allow us to reuse the model in the future:

```
dump(km, '../data/kmeans.joblib')
km_ = load('../data/kmeans.joblib')
prediction = km_.predict(vectorized)
print(prediction)
```

The result might look like this:

```
[0]
```

Using SVMs for supervised text classification

In this recipe, we will build a machine learning classifier that uses the SVM algorithm. By the end of this recipe, you will have a working classifier that you will be able to test on new inputs and evaluate using the same `classification_report` tools we used in the previous sections. We will use the same BBC news dataset we used with KMeans previously.

Getting ready

We will continue working with the same packages that we already installed in the previous recipes. The packages needed are installed in the `poetry` environment or by installing the `requirements.txt` file.

The notebook is located at `https://github.com/PacktPublishing/Python-Natural-Language-Processing-Cookbook-Second-Edition/blob/main/Chapter04/4.4-svm_classification.ipynb`.

How to do it...

We will load the cleaned training and test data that we had saved in the previous recipe. We will then create the SVM classifier and train it. We will use BERT encoding as our vectorizer.

Your steps should be formatted like so:

1. Run the simple classifier file:

    ```
    %run -i "../util/util_simple_classifier.ipynb"
    ```

2. Import the necessary functions and packages:

    ```
    from sklearn.svm import SVC
    from sentence_transformers import SentenceTransformer
    from sklearn.metrics import confusion_matrix
    ```

3. Here, we load the training and test data. If you get a `FileNotFoundError` error in this step, run steps 1-7 from the previous recipe, *Clustering sentences using K-Means – unsupervised text classification*. We then shuffle the training data using the `sample` function. Shuffling ensures that we do not have long sequences of data of the same class. Finally, we print out the number of counts of examples by class. We see that the classes are more or less balanced, which is important for training a classifier:

    ```
    train_df = pd.read_json("../data/bbc_train.json")
    test_df = pd.read_json("../data/bbc_test.json")
    train_df.sample(frac=1)
    print(train_df.groupby('label_text').count())
    print(test_df.groupby('label_text').count())
    ```

    The result will look like this:

	text	label	text_tokenized	text_clean	cluster
label_text					
business	231	231	231	231	231
entertainment	181	181	181	181	181
politics	182	182	182	182	182
sport	243	243	243	243	243
tech	194	194	194	194	194
	text	label	text_tokenized	text_clean	
label_text					
business	58	58	58	58	
entertainment	45	45	45	45	
politics	45	45	45	45	
sport	61	61	61	61	
tech	49	49	49	49	

4. Here, we load the sentence transformer `all-MiniLM-L6-v2` model that will provide the vectors for us. To learn more about the model, please read the *Using BERT and OpenAI embeddings instead of word embeddings* recipe in *Chapter 3*. We then define the `get_sentence_vector` function, which returns the sentence embedding for the text input:

```
model = SentenceTransformer('all-MiniLM-L6-v2')
def get_sentence_vector(text, model):
    sentence_embeddings = model.encode([text])
    return sentence_embeddings[0]
```

5. Define a function that will create an SVM object and train it given input data. It takes in the input vectors and the gold labels, creates an SVC object with the RBF kernel and a regularization parameter of 0.1, and trains it on the training data. It then returns the trained classifier:

```
def train_classifier(X_train, y_train):
    clf = SVC(C=0.1, kernel='rbf')
    clf = clf.fit(X_train, y_train)
    return clf
```

6. In this step, we create the list of labels for the classifier and the `vectorize` method. We then create the training and test datasets using the `create_train_test_data` method, which is located in the simple classifier file. We then train the classifier using the `train_classifier` function and print the training and test metrics. We see that the test metrics are really good, all above 90%:

```
target_names=["tech", "business", "sport",
    "entertainment", "politics"]
vectorize = lambda x: get_sentence_vector(x, model)
(X_train, X_test, y_train, y_test) = create_train_test_data(
    train_df, test_df, vectorize, column_name="text_clean")
clf = train_classifier(X_train, y_train)
print(classification_report(train_df["label"],
        y_train, target_names=target_names))
test_classifier(test_df, clf, target_names=target_names)
```

The output will be as follows:

```
                precision    recall  f1-score   support

          tech       1.00      1.00      1.00       194
      business       1.00      1.00      1.00       231
         sport       1.00      1.00      1.00       243
 entertainment       1.00      1.00      1.00       181
      politics       1.00      1.00      1.00       182

      accuracy                           1.00      1031
```

	precision	recall	f1-score	support
macro avg	1.00	1.00	1.00	1031
weighted avg	1.00	1.00	1.00	1031

	precision	recall	f1-score	support
tech	0.92	0.98	0.95	49
business	0.95	0.90	0.92	58
sport	1.00	1.00	1.00	61
entertainment	1.00	0.98	0.99	45
politics	0.96	0.98	0.97	45
accuracy			0.97	258
macro avg	0.97	0.97	0.97	258
weighted avg	0.97	0.97	0.96	258

7. In this step, we print out the confusion matrix to see where the classifier makes mistakes. The rows represent the correct labels, and the columns are the predicted labels. We see the most confusion (four examples) where the correct label is business but tech is predicted, and where business is the correct label and politics is predicted (two examples). We also see that business is predicted incorrectly for tech, entertainment, and politics once each. These errors are also reflected in the metrics, where we see that both recall and precision for business are affected. The only category with perfect scores is sport and it also has zeroes across the confusion matrix everywhere except the intersection of the correct row and predicted column. We can use the confusion matrix to see which categories have the most confusion between themselves and take measures to rectify that if needed:

```
print(confusion_matrix(test_df["label"], test_df["prediction"]))
[[48  1  0  0  0]
 [ 4 52  0  0  2]
 [ 0  0 61  0  0]
 [ 0  1  0 44  0]
 [ 0  1  0  0 44]]
```

8. We will test the classifier on a new example. We first vectorize the text and then use the trained model to make a prediction and print the prediction. The new article is about tech, and the prediction is class 0, which is indeed tech:

```
new_example = """iPhone 12: Apple makes jump to 5G
Apple has confirmed its iPhone 12 handsets will be its first to
work on faster 5G networks.
The company has also extended the range to include a new "Mini"
model that has a smaller 5.4in screen.
The US firm bucked a wider industry downturn by increasing its
handset sales over the past year.
But some experts say the new features give Apple its best
```

```
opportunity for growth since 2014, when it revamped its line-up
with the iPhone 6.
"5G will bring a new level of performance for downloads and
uploads, higher quality video streaming, more responsive gaming,
real-time interactivity and so much more," said chief executive
Tim Cook.
..."""
vector = vectorize(new_example)
prediction = clf.predict([vector])
print(prediction))
```

The result will be as follows:

```
[0]
```

There's more...

There are many different machine learning algorithms that can be used instead of the SVM algorithm. Some of the others include regression, Naïve Bayes, and decision trees. You can experiment with them and see which ones perform better.

Training a spaCy model for supervised text classification

In this recipe, we will train a spaCy model on the BBC dataset, the same dataset we used in the previous recipe, to will predict the text category.

Getting ready

We will use the spaCy package to train our model. All the dependencies are taken care of by the poetry environment.

You will need to download the config file from the book's GitHub repository, located at https://github.com/PacktPublishing/Python-Natural-Language-Processing-Cookbook-Second-Edition/blob/main/data/spacy_config.cfg. This file should be located at the path ../data/spacy_config.cfg with respect to the notebook.

> **Note**
> You can modify the training config, or generate your own at https://spacy.io/usage/training.

The notebook is located at https://github.com/PacktPublishing/Python-Natural-Language-Processing-Cookbook-Second-Edition/blob/main/Chapter04/4.5-spacy_textcat.ipynb.

How to do it...

The general structure of the training is similar to a plain machine learning model training, where we clean the data, create the dataset, and split it into training and testing datasets. We then train a model and test it on unseen data:

1. Run the simple classifier file:

    ```
    %run -i "../util/lang_utils.ipynb"
    ```

2. Import the necessary functions and packages:

    ```python
    import pandas as pd
    from spacy.cli.train import train
    from spacy.cli.evaluate import evaluate
    from spacy.cli.debug_data import debug_data
    from spacy.tokens import DocBin
    ```

3. Here we define the `preprocess_data_entry` function, which will take the input text, its label, and the list of all labels. It will then run the small spaCy model on the text. This model was imported by running the language utilities file in step 1. It is not important which model we use in this step, since we just want to have a `Doc` object created from the text. That is why we run the smallest model, so it takes less time. We then create a one-hot encoding for the text class, setting the class label to 1 and the rest to 0. We then create a label dictionary that maps the category name to its value. We set the `doc.cats` attribute to this dictionary and return the `Doc` object. spaCy requires this preprocessing of the data to train a classification model:

    ```python
    def preprocess_data_entry(input_text, label, label_list):
        doc = small_model(input_text)
        cats = [0] * len(label_list)
        cats[label] = 1
        final_cats = {}
        for i, label in enumerate(label_list):
            final_cats[label] = cats[i]
        doc.cats = final_cats
        return doc
    ```

4. Now we prepare the training and test datasets. We create the `DocBin` objects for both training and test data that is required by the spaCy algorithm. We then load the saved data from disk. This is the data we saved in the K-Means recipe. If you get a `FileNotFoundError` error here, you need to run steps 1-7 from the *Clustering sentences using K-Means – unsupervised text classification* recipe. We then shuffle the training dataframe. Then we preprocess each data point using the function we defined in the previous step. We then add each datapoint to the `DocBin` object. Finally, we save the two datasets to disk:

    ```python
    train_db = DocBin()
    test_db = DocBin()
    ```

```
label_list = ["tech", "business", "sport",
    "entertainment", "politics"]
train_df = pd.read_json("../data/bbc_train.json")
test_df = pd.read_json("../data/bbc_test.json")
train_df.sample(frac=1)
for idx, row in train_df.iterrows():
    text = row["text"]
    label = row["label"]
    doc = preprocess_data_entry(text, label, label_list)
    train_db.add(doc)
for idx, row in test_df.iterrows():
    text = row["text"]
    label = row["label"]
    doc = preprocess_data_entry(text, label, label_list)
    test_db.add(doc)
train_db.to_disk('../data/bbc_train.spacy')
test_db.to_disk('../data/bbc_test.spacy')
```

5. Train the model using the `train` command. In order for training to work, you will need to have the configuration file downloaded to the `data` folder. This is explained in the *Getting ready* section of this recipe. The training config specifies the location of the training and test datasets, so you need to run the previous step for the training to work. The `train` command saves the model in the `model_last` subdirectory of the directory we specify in the input (`../models/spacy_textcat_bbc/` in this case):

```
train("../data/spacy_config.cfg", output_path="../models/spacy_
textcat_bbc")
```

The output will differ but might look like this (truncated for easier reading). We see that the final accuracy of our trained model is 85%:

```
i Saving to output directory: ../models/spacy_textcat_bbc
i Using CPU


============================ Initializing pipeline
============================
✔ Initialized pipeline
4.5-spacy_textcat.ipynb
============================= Training pipeline
============================
i Pipeline: ['tok2vec', 'textcat']
i Initial learn rate: 0.001
E    #        LOSS TOK2VEC  LOSS TEXTCAT  CATS_SCORE  SCORE
---  ------   ------------  ------------  ----------  ------
  0    0           0.00          0.16        8.48      0.08
  0  200          20.77         37.26       35.58      0.36
```

```
  0     400        98.56     35.96     26.90    0.27
  0     600        49.83     37.31     36.60    0.37
... (truncated)
  4    4800      7571.47      9.64     80.25    0.80
  4    5000     16164.99     10.58     87.71    0.88
  5    5200      8604.43      8.20     84.98    0.85
✔ Saved pipeline to output directory
../models/spacy_textcat_bbc/model-last
```

6. Now we test the model on an unseen example. We first load the model and then get an example from the test data. We then check the text and its category. We run the model on the input text and print the resulting probabilities. The model will give a dictionary of categories with their respective probability scores. These scores indicate the probability that the text belongs to the respective class. The class with the highest probability is the one we should assign to the text. The category dictionary is in the doc.cats attribute, just like when we were preparing the data, but in this case the model assigns it. In this case, the text is about politics and the model correctly classifies it:

```
nlp = spacy.load("../models/spacy_textcat_bbc/model-last")
input_text = test_df.iloc[1, test_df.columns.get_loc('text')]
print(input_text)
print(test_df["label_text"].iloc[[1]])
doc = nlp(input_text)
print("Predicted probabilities: ", doc.cats)
```

The output will look similar to this:

```
lib dems  new election pr chief the lib dems have appointed
a senior figure from bt to be the party s new communications
chief for their next general election effort.  sandy walkington
will now work with senior figures such as matthew taylor on
completing the party manifesto. party chief executive lord
rennard said the appointment was a  significant strengthening
of the lib dem team . mr walkington said he wanted the party
to be ready for any  mischief  rivals or the media tried to
throw at it.   my role will be to ensure this new public profile
is effectively communicated at all levels   he said.  i also
know the party will be put under scrutiny in the media and
from the other parties as never before - and we will need to
show ourselves ready and prepared to counter the mischief and
misrepresentation that all too often comes from the party s
opponents.   the party is already demonstrating on every issue
that it is the effective opposition.   mr walkington s new job
title is director of general election communications.
8     politics
Name: label_text, dtype: object
Predicted probabilities:  {'tech': 3.531841841208916e-
08, 'business': 0.000641813559923321, 'sport':
0.00033847044687718153, 'entertainment': 0.0001617442321730776,
'politics': 0.9988579750061035}
```

7. In this step, we define a `get_prediction` function, which takes text, a spaCy model, and the list of potential classes and outputs the category whose probability is the highest. We then apply this function to the `text` column of the test dataframe:

```
def get_prediction(input_text, nlp_model, target_names):
    doc = nlp_model(input_text)
    category = max(doc.cats, key = doc.cats.get)
    return target_names.index(category)
test_df["prediction"] = test_df["text"].apply(
    lambda x: get_prediction(x, nlp, label_list))
```

8. Now we print out the classification report based on the data from the test dataframe we generated in the previous step. The overall accuracy of the model is 87%, and the reason it is a bit low is because we do not have enough data to train a better model:

```
print(classification_report(test_df["label"],
    test_df["prediction"], target_names=target_names))
```

The result should look similar to this:

	precision	recall	f1-score	support
tech	0.82	0.94	0.87	80
business	0.94	0.83	0.89	102
sport	0.89	0.89	0.89	102
entertainment	0.94	0.87	0.91	77
politics	0.78	0.83	0.80	84
accuracy			0.87	445
macro avg	0.87	0.87	0.87	445
weighted avg	0.88	0.87	0.87	445

9. In this step, we do the same evaluation using the spaCy `evaluate` command. This command takes in the path to the model and the path to the test dataset and outputs the scores in a slightly different format. We see that the scores from both steps are consistent:

```
evaluate('../models/spacy_textcat_bbc/model-last', '../data/
bbc_test.spacy')
```

The result should look similar to this:

```
{'token_acc': 1.0,
 'token_p': 1.0,
 'token_r': 1.0,
 'token_f': 1.0,
 'cats_score': 0.8719339318444819,
 'cats_score_desc': 'macro F',
 'cats_micro_p': 0.8719101123595505,
```

```
  'cats_micro_r': 0.8719101123595505,
  'cats_micro_f': 0.8719101123595505,
  'cats_macro_p': 0.8746516896205309,
  'cats_macro_r': 0.8732906799083269,
  'cats_macro_f': 0.8719339318444819,
  'cats_macro_auc': 0.9800144873453936,
  'cats_f_per_type': {'tech': {'p': 0.8152173913043478,
    'r': 0.9375,
    'f': 0.872093023255814},
   'business': {'p': 0.9444444444444444,
    'r': 0.8333333333333334,
    'f': 0.8854166666666667},
   'sport': {'p': 0.8921568627450981,
    'r': 0.8921568627450981,
    'f': 0.8921568627450981},
   'entertainment': {'p': 0.9436619718309859,
    'r': 0.8701298701298701,
    'f': 0.9054054054054054},
   'politics': {'p': 0.7777777777777778,
    'r': 0.8333333333333334,
    'f': 0.8045977011494253}},
  'cats_auc_per_type': {'tech': 0.9842808219178081,
   'business': 0.9824501229063054,
   'sport': 0.9933544846510032,
   'entertainment': 0.9834839073969509,
   'politics': 0.9565030998549005},
  'speed': 6894.989948433934}
```

Classifying texts using OpenAI models

In this recipe, we will ask an OpenAI model to provide the classification of an input text. We will use the same BBC dataset from previous recipes.

Getting ready

To run this recipe, you will need to have the openai package installed, provided as part of the poetry environment, and the requirements.txt file. You will also have to have an OpenAI API key. Paste it into the provided field in the file utilities notebook (https://github.com/PacktPublishing/Python-Natural-Language-Processing-Cookbook-Second-Edition/blob/main/util/file_utils.ipynb).

The notebook is located at https://github.com/PacktPublishing/Python-Natural-Language-Processing-Cookbook-Second-Edition/blob/main/Chapter04/4.6_openai_classification.ipynb.

> **Note**
> OpenAI frequently changes and retires existing models and introduces new ones. The model we use in this recipe, `gpt-3.5-turbo`, might be obsolete by the time you read this. In this case, please check the OpenAI documentation and select another suitable model.

How to do it...

In this recipe, we will query the OpenAI API and provide a request for classification as the prompt. We will then post-process the results and evaluate the Open AI model on this task:

1. Run the simple classifier and the file utilities notebooks:

```
%run -i "../util/file_utils.ipynb"
%run -i "../util/util_simple_classifier.ipynb"
```

2. Import the necessary functions and packages to create the OpenAI client using the API key:

```
import re
from sklearn.metrics import classification_report
from openai import OpenAI
client = OpenAI(api_key=OPEN_AI_KEY)
```

3. Load the training and test datasets using Hugging Face without preprocessing them for the number of classes, as we will not be training a new model:

```
train_dataset = load_dataset("SetFit/bbc-news", split="train")
test_dataset = load_dataset("SetFit/bbc-news", split="test")
```

4. Load and print the first example in the dataset and its category:

```
example = test_dataset[0]["text"]
category = test_dataset[0]["label_text"]
print(example)
print(category)
```

The result should be as follows:

```
carry on star patsy rowlands dies actress patsy rowlands  known
to millions for her roles in the carry on films  has died at
the age of 71.  rowlands starred in nine of the popular carry
on films  alongside fellow regulars sid james  kenneth williams
and barbara windsor. she also carved out a successful television
career  appearing for many years in itv s well-loved comedy
bless this house....
entertainment
```

5. Run the OpenAI model on this one example. In step 5, we query the OpenAI API asking it to classify this example. We create the prompt and append the example text to it. In the prompt, we specify to the model that it is to classify the input text as one of five classes and the output

format. If we don't include these output instructions, it might add other words to it and return text such as *The topic is entertainment*. We select the gpt-3.5-turbo model and specify the prompt, the temperature, and several other parameters. We set the temperature to 0 so that there is no or minimal variation in the model's response. We then print the response returned by the API. The output might vary, but in most cases, it should return *entertainment*, which is correct:

```
prompt="""You are classifying texts by topics. There are 5
topics: tech, entertainment, business, politics and sport.
Output the topic and nothing else. For example, if the topic is
business, your output should be "business".
Give the following text, what is its topic from the above list
without any additional explanations: """ + example
response = client.chat.completions.create(
    model="gpt-3.5-turbo",
    temperature=0,
    max_tokens=256,
    top_p=1.0,
    frequency_penalty=0,
    presence_penalty=0,
    messages=[
        {"role": "system", "content":
            "You are a helpful assistant."},
        {"role": "user", "content": prompt}
    ],
)
print(response.choices[0].message.content)
```

The result might vary, but should look like this:

```
entertainment
```

6. Create a function that will provide the classification of an input text and return the category. It takes input text and calls the OpenAI API with the same prompt we used previously. It then lowercases the response, strips it of extra white space, and returns it:

```
def get_gpt_classification(input_text):
    prompt="""You are classifying texts by topics. There are 5
topics: tech, entertainment, business, politics and sport.
Output the topic and nothing else. For example, if the topic is
business, your output should be "business".
Give the following text, what is its topic from the above list
without any additional explanations: """ + input_text
    response = client.chat.completions.create(
        model="gpt-3.5-turbo",
        temperature=0,
        max_tokens=256,
        top_p=1.0,
        frequency_penalty=0,
```

```
            presence_penalty=0,
            messages=[
                {"role": "system", "content":
                    "You are a helpful assistant."},
                {"role": "user", "content": prompt}
            ],
    )
    classification = response.choices[0].message.content
    classification = classification.lower().strip()
    return classification
```

7. In this step, we load test data. We take the test dataset from Hugging Face and convert it into a dataframe. We then shuffle the dataframe and select the first 200 examples. The reason is that we want to reduce the cost of testing this classifier through the OpenAI API. You can modify how much data you test this method on:

```
test_df = test_dataset.to_pandas()
test_df.sample(frac=1)
test_data = test_df[0:200].copy()
```

8. In step 8, we use the `get_gpt_classification` function to create a new column in the test dataframe. Depending on the number of test examples you have, it might take a few minutes to run:

```
test_data["gpt_prediction"] = test_data["text"].apply(
    lambda x: get_gpt_classification(x))
```

9. Despite our instructions to OpenAI to only provide the category as the answer, it might add some other words, so we define a function, `get_one_word_match`, that cleans OpenAI's output. In this function, we use a regular expression to match one of the class labels and return just that word from the original string. We then apply this function to the `gpt_prediction` column in the test dataframe:

```
def get_one_word_match(input_text):
    loc = re.search(
        r'tech|entertainment|business|sport|politics',
        input_text).span()
    return input_text[loc[0]:loc[1]]
test_data["gpt_prediction"] = test_data["gpt_prediction"].apply(
    lambda x: get_one_word_match(x))
```

10. Now we turn the label into numerical format:

```
label_list = ["tech", "business", "sport",
    "entertainment", "politics"]
test_data["gpt_label"] = test_data["gpt_prediction"].apply(
    lambda x: label_list.index(x))
```

11. We print the resulting dataframe. We can see that we have all the information we need to perform an evaluation. We have both the correct labels (the `label` column) and the predicted labels (the `gpt_label` column):

```
print(test_data)
```

The result should look similar to this:

```
                                                      text   label
       label_text    \
0      carry on star patsy rowlands dies actress pats...       3
    entertainment
1      sydney to host north v south game sydney will ...       2
            sport
..                                                     ...    ...
                 ...
198    xbox power cable  fire fear  microsoft has sai...       0
             tech
199    prop jones ready for hard graft adam jones say...       2
            sport


     gpt_prediction  gpt_label
0     entertainment          3
1             sport          2
..              ...        ...
198            tech          0
199           sport          2
```

12. Now we can print the classification report that evaluates the OpenAI classification:

```
print(classification_report(test_data["label"],
        test_data["gpt_label"], target_names=label_list))
```

The results might vary. This is a sample output. We see that the overall accuracy is good, 90%:

```
                 precision    recall  f1-score   support

          tech       0.97      0.80      0.88        41
      business       0.87      0.89      0.88        44
         sport       1.00      0.96      0.98        48
 entertainment       0.88      0.90      0.89        40
      politics       0.76      0.96      0.85        27

      accuracy                           0.90       200
     macro avg       0.90      0.90      0.90       200
  weighted avg       0.91      0.90      0.90       200
```

5

Getting Started with Information Extraction

In this chapter, we will cover the basics of **information extraction**. Information extraction is the task of pulling very specific information from text. For example, you might want to know the companies mentioned in a news article. Instead of spending time reading the whole article, you can use information extraction techniques to access the companies almost instantly.

We will start with extracting emails addresses and URLs from job announcements. Then, we will use an algorithm called **Levenshtein distance** to find similar strings. Next, we will extract important keywords from text. After that, we will use **spaCy** to find named entities in text, and later, we will train our own named entity recognition model in spaCy. We will then do basic sentiment analysis, and, finally, we will train two custom sentiment analysis models.

You will learn how to use existing tools and train your own models for information extraction tasks.

We will cover the following recipes in this chapter:

- Using regular expressions
- Finding similar strings – Levenshtein distance
- Extracting keywords
- Performing named entity recognition using spaCy
- Training your own NER model with spaCy
- Fine-tuning BERT for NER

Technical requirements

The code for this chapter is in a folder named `Chapter05` in the GitHub repository of the book (`https://github.com/PacktPublishing/Python-Natural-Language-Processing-Cookbook-Second-Edition/tree/main/Chapter05`).

As in previous chapters, the packages required for this chapter are part of the Poetry environment. Alternatively, you can install all the packages using the `requirements.txt` file.

Using regular expressions

In this recipe, we will use regular expressions to find email addresses and URLs in text. Regular expressions are special character sequences that define search patterns and can be created and used via the Python `re` package. We will use a job descriptions dataset and write two regular expressions, one for emails and one for URLs.

Getting ready

Download the job descriptions dataset here: `https://www.kaggle.com/andrewmvd/data-scientist-jobs`. It is also available in the book's GitHub repository at `https://github.com/PacktPublishing/Python-Natural-Language-Processing-Cookbook-Second-Edition/blob/main/data/DataScientist.csv`. Save it into the `/data` folder.

The notebook is located at `https://github.com/PacktPublishing/Python-Natural-Language-Processing-Cookbook-Second-Edition/blob/main/Chapter05/5.1_regex.ipynb`.

How to do it...

We will read the data from the CSV file into a `pandas` DataFrame and will use the Python `re` package to create regular expressions and search the text. The steps are as follows:

1. Import the `re` and `pandas` packages:

    ```
    import re
    import pandas as pd
    ```

2. Read in the data and check the contents inside it:

    ```
    data_file = "../data/DataScientist.csv"
    df = pd.read_csv(data_file, encoding='utf-8')
    print(df)
    ```

The output will be long and should start like this:

```
      Unnamed: 0  index                                  Job Title  \
0              0      0                     Senior Data Scientist
1              1      1            Data Scientist, Product Analytics
2              2      2                     Data Science Manager
3              3      3                              Data Analyst
4              4      4                     Director, Data Science
...          ...    ...                                       ...
3904        3904   4375                         AWS Data Engineer
3905        3905   4376                    Data Analyst â Junior
3906        3906   4377          Security Analytics Data Engineer
3907        3907   4378          Security Analytics Data Engineer
3908        3908   4379  Patient Safety Physician or Safety Scientist -...
```

Figure 5.1 – DataFrame output

3. The get_list_of_items helper function takes a DataFrame as input and turns one of its columns into a list. It accepts the DataFrame and the column name as inputs. First, it gets the column values, which is a list of lists, and then flattens that list. It then removes duplicates by turning the list into a set and casts it back to a list:

```
def get_list_of_items(df, column_name):
    values = df[column_name].values
    values = [item for sublist in values for item in sublist]
    list_of_items = list(set(values))
    return list_of_items
```

4. In this step, we define the get_emails function to get all the emails that appear in the Job Description column. The regular expression consists of three parts that appear in square brackets followed by quantifiers:

   * [^\s:|()\']+ is the username part of the regular expression, followed by the @ sign. It consists of one group of characters, which is shown in square brackets. Any characters from this group may appear in the username one or more times. This is shown using the + quantifier. The characters in the username can be anything but a space (\s), colon (:), pipe (|), and apostrophe ('). The ^ character shows the negation of the character class. An apostrophe is a special character in regular expressions and has to be escaped with a backward slash in order to invoke the regular meaning of the character.

   * [a-zA-Z0-9\.]+ is the first part of the domain name, followed by a dot. This part is simply alphanumeric characters, lowercase or uppercase, and a dot appearing one or more times. Since a dot is a special character, we escape it with a backward slash. The a-z expression signifies a range of characters from *a* to *z*.

- [a-zA-Z] + is the last part of the domain name, which is the top-level domain, such as .com, .org, and so on. Usually, no digits are allowed in these top-level domains, and the regular expression matches lowercase or uppercase characters that appear one or more times.

This regular expression is sufficient to parse all emails in the dataset and not present any false positives. You might find that, in your data, there are additional adjustments that need to be made to the regular expression:

```
def get_emails(df):
    email_regex = '[^\s:|()\']+@[a-zA-Z0-9\.]+\.[a-zA-Z]+'
    df['emails'] = df['Job Description'].apply(
        lambda x: re.findall(email_regex, x))
    emails = get_list_of_items(df, 'emails')
    return emails
```

5. We will now get the emails from the DataFrame using the previous functions:

```
emails = get_emails(df)
print(emails)
```

Part of the result will look like this:

```
['hrhelpdesk@phila.gov', 'talent@quartethealth.com', …,
'careers@edo.com', 'Talent.manager@techquarry.com', 'resumes@
nextgentechinc.com', …, 'talent@ebay.com', …, 'info@springml.
com',…]
```

6. The get_urls helper function takes a DataFrame as input and turns one of its columns into a list.

URLs are significantly more complicated than emails. Here is the breakdown of the regular expression:

- http[s]?://: This is the http part of the URL. All URLs in this dataset had this part, but that might not be the case in your data and you will have to adjust the regular expression accordingly. This part of the expression will match both http and https, since s is listed as appearing zero or one time, which is signified by the ? quantifier.

- (www\.)?: Next, we have a *group* of characters, which are treated as a unit, but all have to appear in the order in which they are listed. In this case, this is the www part of the URL, followed by a dot, escaped with a backward slash. The group of characters may appear zero or one time, signified by the ? character at the end.

- [A-Za-z0-9-_\.\-]+: This part is the domain name of the website, followed by the top-level domain. Website names also include dashes, and the dot character appears before the top-level domain and subdomains.

- /? [A-Za-z0-9$\-_\-\/\.\?]*): This part is whatever follows the domain name after the slash. It could be a variety of characters that list files, parameters, and so on. They could or could not be present, and that is why they are followed by the * quantifier. The bracket at the end signifies the end of the matching group.

- [\.)\"]*: Many URLs in this dataset are followed by dots, brackets, and other characters, and this is the last part of the regular expression.

In this function, we use the `finditer` function from the `re` package. It finds all matches in a text and returns them as `Match` objects. We can find the start and end of the match by using the `span()` object method. It returns a tuple, where the first element is the start and the second element is the end of the match:

```
def get_urls(df):
    url_regex = '(http[s]?://(www\.)?[A-Za-z0-9-_\.\-]+\.[A-Za-z]+/?[A-Za-z0-9$\-_\-\/\.]*)[\.)\"]*'
    df['urls'] = df['Job Description'].apply(
        lambda x: [
            x[item.span()[0]:item.span()[1]]
            for item in re.finditer(url_regex, x)
        ]
    )
    urls = get_list_of_items(df, 'urls')
    return urls
```

7. We will get the URLs in a similar fashion:

```
urls = get_urls(df)
print(urls)
```

Part of the result will look like this:

```
['https://youtu.be/c5TgbpE9UBI', 'https://www.linkedin.
com/in/emma-riley-72028917a/', 'https://www.dol.gov/
ofccp/regs/compliance/posters/ofccpost.htm', 'https://
www.naspovaluepoint.org/portfolio/mmis-provider-services-
module-2018-2028/hhs-technology-group/).', 'https://www.
instagram.com/gatestonebpo', 'http://jobs.sdsu.edu', 'http://
www.colgatepalmolive.com.', 'http://www1.eeoc.gov/employers/
upload/eeoc_self_print_poster.pdf', 'https://www.gofundme.
com/2019https', 'https://www.decode-m.com/', 'https://bit.
ly/2lCOcYS',…]
```

There's more...

Writing regular expressions can quickly turn into a messy affair. I use regular expression testing websites to enter the text in which I expect a match and the regular expression. One example of such a site is https://regex101.com/.

Finding similar strings – Levenshtein distance

When doing information extraction, in many cases, we deal with misspellings, which can bring complications to the task. To get around this problem, several methods are available, including Levenshtein distance. This algorithm finds the number of edits/additions/deletions needed to change one string into another. For example, to change the word *put* into *pat*, you need to substitute *u* for *a*, and that is one change. To change the word *kitten* into *smitten*, you need to do two edits: change *k* into *m* and add an *s* at the start.

In this recipe, you will be able to use this technique to find a match to a misspelled email.

Getting ready

We will use the same packages and the data scientist job description dataset that we used in the previous recipe, and the `python-Levenshtein` package, which is part of the Poetry environment and is included in the `requirements.txt` file.

The notebook is located at `https://github.com/PacktPublishing/Python-Natural-Language-Processing-Cookbook-Second-Edition/blob/main/Chapter05/5.2_similar_strings.ipynb`.

How to do it...

We will read the dataset into a `pandas` DataFrame and use the emails extracted from it to search for a misspelled email. Your steps should be formatted like so:

1. Run the language utilities file. This file contains the `get_emails` function we created in the previous recipe:

    ```
    %run -i "../util/lang_utils.ipynb"
    ```

2. Do the necessary imports:

    ```
    import pandas as pd
    import Levenshtein
    ```

3. Read the data into a `pandas` DataFrame object:

    ```
    data_file = "../data/DataScientist.csv"
    df = pd.read_csv(data_file, encoding='utf-8')
    ```

4. Filter out all emails from the DataFrame using the `get_emails` function, which is explained in more detail in the previous recipe, *Using regular expressions*:

    ```
    emails = get_emails(df)
    ```

5. The `find_levenshtein` function takes in a DataFrame and an input string and computes the Levenshtein distance between it and each string in the emails column. It takes in an input string and a DataFrame with emails and creates a new column in which the value is the Levenshtein distance between the input and the email address in the DataFrame. The column name is `distance_to_[input_string]`:

```
def find_levenshtein(input_string, df):
    df['distance_to_' + input_string] = \
        df['emails'].apply(lambda x: Levenshtein.distance(
            input_string, x))
    return df
```

6. In this step, we define the `get_closest_email_lev` function, which takes in a DataFrame with emails and an email to match and returns the email in the DataFrame that is closest to the input. We accomplish this by using the `find_levenshtein` function to create a new column with distances to the input email and then using the `idxmin()` function from `pandas` to find the index of the minimum value. We use the minimum index to find the closest email:

```
def get_closest_email_lev(df, email):
    df = find_levenshtein(email, df)
    column_name = 'distance_to_' + email
    minimum_value_email_index = df[column_name].idxmin()
    email = df.loc[minimum_value_email_index]['emails']
    return email
```

7. Next, we load the emails into a new DataFrame and use the misspelled email address `rohitt.macdonald@prelim.com` to find a match in the new `email` DataFrame:

```
new_df = pd.DataFrame(emails,columns=['emails'])
input_string = "rohitt.macdonald@prelim.com"
email = get_closest_email_lev(new_df, input_string)
print(email)
```

The function returns `rohit.mcdonald@prolim.com`, the correct spelling of the email address:

```
rohit.mcdonald@prolim.com
```

There's more…

The Levenshtein package includes other string similarity measuring methods, which you can explore at `https://rapidfuzz.github.io/Levenshtein/`. In this section, we look at the **Jaro distance**.

We can use another function, the Jaro similarity, which outputs similarity between two strings as a number between 0 and 1, where 1 means that two strings are the same. The process is similar, but we need the index with the maximum value instead of the minimum since the Jaro similarity function returns a higher value for more similar strings. Let's go through the steps:

1. The `find_jaro` function takes in a DataFrame and an input string and computes the Jaro similarity between it and each string in the email column:

    ```
    def find_jaro(input_string, df):
        df['distance_to_' + input_string] = df['emails'].apply(
            lambda x: Levenshtein.jaro(input_string, x)
        )
        return df
    ```

2. The `get_closest_email_jaro` function uses the function we defined in the previous step to find the email address that is closest to the one input:

    ```
    def get_closest_email_jaro(df, email):
        df = find_jaro(email, df)
        column_name = 'distance_to_' + email
        maximum_value_email_index = df[column_name].idxmax()
        email = df.loc[maximum_value_email_index]['emails']
        return email
    ```

3. Next, we use the misspelled email address `rohitt.macdonald@prelim.com` to find a match in the new email DataFrame:

    ```
    email = get_closest_email_jaro(new_df, input_string)
    print(email)
    ```

    The output is as follows:

    ```
    rohit.mcdonald@prolim.com
    ```

4. An extension of the Jaro similarity function is the **Jaro-Winkler function**, which attaches a weight to the end of the word, and that weight lowers the importance of misspellings toward the end. For example, let's look at the following function:

    ```
    print(Levenshtein.jaro_winkler("rohit.mcdonald@prolim.com",
        "rohit.mcdonald@prolim.org"))
    ```

    This outputs the following:

    ```
    1.0
    ```

Extracting keywords

In this recipe, we will extract keywords from a text. We will be working with the BBC news dataset that contains news articles. You can learn more about the dataset in *Chapter 4*, in the recipe titled *Clustering sentences using K-Means: unsupervised text classification*.

Extracting keywords from text can give us a quick idea about what the article is about and can also serve as a basis for a tagging system, for example, on a website.

For the extraction to work correctly, we need to train a TF-IDF vectorizer that we will use during the extraction phase.

Getting ready

In this recipe, we will use the `sklearn` package. It is part of the Poetry environment. You can also install it together with other packages by installing the `requirements.txt` file.

The BBC news dataset is available on Hugging Face at `https://huggingface.co/datasets/SetFit/bbc-news`.

The notebook is located at `https://github.com/PacktPublishing/Python-Natural-Language-Processing-Cookbook-Second-Edition/blob/main/Chapter05/5.3_keyword_extraction.ipynb`.

How to do it...

To extract keywords from a given text, we first need a corpus of text that we will fit the vectorizer on. Once that is done, we can use it to extract keywords from a text that is similar to the processed corpus. Here are the steps:

1. Run the language utilities notebook:

   ```
   %run -i "../util/lang_utils.ipynb"
   ```

2. Import the necessary packages and functions:

   ```
   from datasets import load_dataset
   from nltk import word_tokenize
   from math import ceil
   from sklearn.feature_extraction.text import TfidfVectorizer
   from nltk.corpus import stopwords
   ```

3. Load the training and testing datasets, convert them to `pandas` DataFrame objects, and print out the training DataFrame to discover how it looks. The DataFrame has three columns, one for the news article text, one for the label in numeric format, and one for the label text:

   ```
   train_dataset = load_dataset("SetFit/bbc-news", split="train")
   test_dataset = load_dataset("SetFit/bbc-news", split="test")
   train_df = train_dataset.to_pandas()
   test_df = test_dataset.to_pandas()
   print(train_df)
   print(test_df)
   ```

The result should look similar to this:

```
        text  label      label_text
0  wales want rugby league training wales could f... 2   sport
1      china aviation seeks rescue deal scandal-hit
j...  business
...       ...     ...              ...
1223  why few targets are better than many the econo...
1  business
1224  boothroyd calls for lords speaker betty boothr...
4  politics


[1225 rows x 3 columns]

        text  label      label_text
0  carry on star patsy rowlands dies actress pats...
3  entertainment
1      sydney to host north v south game sydney will ... 2   sport
..       ...     ...              ...
998  stormy year for property insurers a string of ...
1  business
999  what the election should really be about  a ge...
4  politics


[1000 rows x 3 columns]
```

4. Create the vectorizer and fit it on the training data text. To learn more about vectorizers, see *Chapter 3*. The TF-IDF vectorizer is discussed in the *Representing texts with TF-IDF* recipe. We use English stopwords, a minimum document frequency of 2, and a maximum document frequency of 95% (to learn more about stopwords, see the *Removing stopwords* recipe in *Chapter 1*):

```
vectorizer = TfidfVectorizer(stop_words='english',
    min_df=2, max_df=0.95)
vectorizer.fit(train_df["text"])
```

5. Now, we will define a few helper functions. The first one will sort a coordinate matrix by the TF-IDF score. It takes the coordinate matrix that is converted from the vector created by the vectorizer. This coordinate matrix's col attribute provides the word indices and the data attribute provides the TF-IDF scores for each word. The function creates a list of tuples from this data, where the first value in the tuple is the index and the second value is the TF-IDF score. It then sorts the tuple list by the TF-IDF score and returns the sorted result. This will give us words that have the maximum TF-IDF score or the ones that are most characteristic of this particular news piece:

```
def sort_data_tfidf_score(coord_matrix):
    tuples = zip(coord_matrix.col, coord_matrix.data)
    return sorted(tuples, key=lambda x: (x[1], x[0]),
        reverse=True)
```

6. The next function, `get_keyword_strings`, will get the keywords for a given vector. It returns the extracted keywords for a given vector. It takes as input the fitted vectorizer, the number of keywords to extract, and the sorted vector of the input text. The function first defines the `index_dict` variable as the dictionary with word indices as keys and corresponding words as values. It then iterates through the sorted vector and appends the words from the dictionary to the `words` list variable. It stops when it reaches the desired number of words. Since the function iterates through the sorted vector, it will give us the words with the highest TF-IDF scores. These words will be the ones most used in this document but not used in other documents, thus giving us an idea about the topic of the article:

```
def get_keyword_strings(vectorizer, num_words, sorted_vector):
    words = []
    index_dict = vectorizer.get_feature_names_out()
    for (item_index, score) in sorted_vector[0:num_words]:
        word = index_dict[item_index]
        words.append(word)
    return words
```

7. The `get_keywords_simple` function will return a list of keywords for a given text. It takes in the input text, the fitted vectorizer, and the desired number of words. It creates a vector for the input text by using the vectorizer, then sorts the vector by using the `sort_data_tfidf_score` function, and finally, gets the top words using the `get_keyword_strings` function:

```
def get_keywords_simple(vectorizer, input_text,
    num_output_words=10):
    vector = vectorizer.transform([input_text])
    sorted = sort_data_tfidf_score(vector.tocoo())
    words = get_keyword_strings(vectorizer, num_output_words,
        sorted)
    return words
```

8. We use the previous function on the first text from the test DataFrame. We take the first article text in the test data and create a list of keywords using the `get_keywords_simple` function. We see that some of the keywords fit the summary, and some are less suitable:

```
print(test_df.iloc[0]["text"])
keywords = get_keywords_simple(vectorizer,
    test_df.iloc[0]["text"])
print(keywords)
```

The result will be as follows:

```
carry on star patsy rowlands dies actress patsy rowlands  known
to millions for her roles in the carry on films  has died at the
age of 71.  rowlands starred in nine of the popular carry on
films  alongside fellow regulars sid james  kenneth williams and
barbara windsor...
['carry', 'theatre', 'scholarship', 'appeared', 'films', 'mrs',
'agent', 'drama', 'died', 'school']
```

There's more...

Now, we will use a more sophisticated approach to extracting keywords from news summaries. We will use a vectorizer that scores not just individual words but also bigrams and trigrams. We will also use the spaCy noun chunks to make sure that the bigrams and trigrams that are output make sense. To learn more about noun chunks, see the *Extracting noun chunks* recipe in *Chapter 2*. The advantage of this method is that we get not only individual words as output but also phrases, such as *Saturday morning* instead of just *Saturday* and *morning* individually.

1. Create the new vectorizer and fit it on the training summaries. We exclude the word the from the stopwords list since spaCy entities might contain it:

```
stop_words = list(stopwords.words('english'))
stop_words.remove("the")
trigram_vectorizer = TfidfVectorizer(
    stop_words=stop_words, min_df=2,
    ngram_range=(1,3), max_df=0.95)
trigram_vectorizer.fit(train_df["summary"])
```

2. Now, define the get_keyword_strings_all function. It will get all the keywords from the sorted vector. It has no restriction on how many words it gets:

```
def get_keyword_strings_all(vectorizer, sorted_vector):
    words = []
    index_dict = vectorizer.get_feature_names_out()
    for (item_index, score) in sorted_vector:
        word = index_dict[item_index]
        words.append(word)
    return words
```

3. Next, we define the get_keywords_complex function that outputs main keywords and phrases up to three words long:

```
def get_keywords_complex(
    vectorizer, input_text, spacy_model, num_words=70
):
    keywords = []
    doc = spacy_model(input_text)
    vector = vectorizer.transform([input_text])
    sorted = sort_coo(vector.tocoo())
    ngrams = get_keyword_strings_all(vectorizer, sorted)
    ents = [ent.text.lower() for ent in doc.noun_chunks]
    for i in range(0, num_words):
        keyword = ngrams[i]
        if keyword.lower() in ents and not
```

```
        keyword.isdigit() and keyword not in keywords:
            keywords.append(keyword)
    return keywords
```

4. Now, we use the previous function on the first test summary:

```
keywords = get_keywords_complex(trigram_vectorizer,
    test_df.iloc[0]["summary"], small_model)
print(keywords)
```

The result will look like this:

```
['the gop', 'the 50 states', 'npr', '11 states', 'state',
'republican governors', 'the dems', 'reelection', 'the helm',
'grabs']
```

Performing named entity recognition using spaCy

Named entity recognition (**NER**) is the task of parsing the names of places, people, organizations, and so on, out of text. This can be useful in many downstream tasks. For example, you could imagine a situation where you would like to sort an article set by the people that are mentioned in it, for example, when carrying out research about a certain person.

In this recipe, we will use NER to parse out named entities from article texts in the BBC dataset. We will load the package and the parsing engine and loop through the NER results.

Getting ready

In this recipe, we will use spaCy. To run it correctly, you will need to download a language model. We will download the small and large models. These models take up significant disk space:

```
python -m spacy download en_core_web_sm
python -m spacy download en_core_web_lg
```

The notebook is located at https://github.com/PacktPublishing/Python-Natural-Language-Processing-Cookbook-Second-Edition/blob/main/Chapter05/5.4_named_entity_extraction.ipynb.

How to do it...

NER happens automatically with the processing that spaCy does for an input text. Accessing the entities happens through the doc.ents variable. We will input an article about Apple's iPhone and see which entities will get parsed from it. Let's see the steps:

1. Run the language utilities file. This will import the necessary packages and functions and initialize the spaCy engine:

```
%run -i "../util/lang_utils.ipynb"
```

2. Initialize the article text. This is an article from `https://www.globalsmt.net/social-media-news/iphone-12-apple-makes-jump-to-5g/`:

```
article = """iPhone 12: Apple makes jump to 5G
Apple has confirmed its iPhone 12 handsets will be its first to
work on faster 5G networks.
The company has also extended the range to include a new "Mini"
model that has a smaller 5.4in screen.
The US firm bucked a wider industry downturn by increasing its
handset sales over the past year.
But some experts say the new features give Apple its best
opportunity for growth since 2014, when it revamped its line-up
with the iPhone 6.
...
"Networks are going to have to offer eye-wateringly attractive
deals, and the way they're going to do that is on great tariffs
and attractive trade-in deals,"
predicted Ben Wood from the consultancy CCS Insight. Apple
typically unveils its new iPhones in September, but opted for a
later date this year.
It has not said why, but it was widely speculated to be related
to disruption caused by the coronavirus pandemic. The firm's
shares ended the day 2.7% lower.
This has been linked to reports that several Chinese internet
platforms opted not to carry the livestream,
although it was still widely viewed and commented on via the
social media network Sina Weibo."""
```

3. Here, we create the spaCy Doc object and use it to extract the entities. The Doc object is created by using the small spaCy model on the text. The model extracts different attributes, including named entities. We print the length of the parsed entities and the entities themselves, together with start and end character information and the entity type (the meaning of the named entity labels can be found in the spaCy documentation at `https://spacy.io/models/en`):

```
doc = small_model(article)
print(len(doc.ents))
small_model_ents = doc.ents
for ent in doc.ents:
    print(ent.text, ent.start_char, ent.end_char, ent.label_)
```

When we print out the result, we see different types of entities, including cardinal numbers, percentages, names of people, dates, organizations, and a NORP entity, which stands for **Nationalities or Religious or Political groups**:

```
44
12 7 9 CARDINAL
Apple 11 16 ORG
5 31 32 CARDINAL
```

```
...
a later date this year 2423 2445 DATE
2.7% 2594 2598 PERCENT
Chinese 2652 2659 NORP
Sina Weibo 2797 2807 PERSON
```

There's more...

We can compare the performance of the small and large models with the following steps:

1. Run the same step as *step 3* from the *How to do it...* section but with the large model:

```
doc = large_model(article)
print(len(doc.ents))
large_model_ents = doc.ents
for ent in doc.ents:
    print(ent.text, ent.start_char, ent.end_char, ent.label_)
```

The result will be as follows:

```
46
12 7 9 CARDINAL
Apple 11 16 ORG
5 31 32 CARDINAL
...
the day 2586 2593 DATE
2.7% 2594 2598 PERCENT
Chinese 2652 2659 NORP
Sina Weibo 2797 2807 PERSON
```

2. There are more entities parsed by the large model, and we can take a look at the differences. We print out two lists; one list contains entities that the small model recognizes, and the other list contains entities that the large model recognizes but not the small:

```
small_model_ents = [str(ent) for ent in small_model_ents]
large_model_ents = [str(ent) for ent in large_model_ents]
in_small_not_in_large = set(small_model_ents) \
    - set(large_model_ents)
in_large_not_in_small = set(large_model_ents) \
    - set(small_model_ents)
print(in_small_not_in_large)
print(in_large_not_in_small)
```

The result will be as follows:

```
{'iPhone 11', 'iPhone', 'iPhones'}
{'6', 'the day', 'IDC', '11', 'Pro', 'G\nApple', 'SE'}
```

You can see that there are some differences between the results provided by the two models.

Training your own NER model with spaCy

In the previous recipe, we used the pretrained spaCy model to extract named entities. This NER model can suffice in many cases. There might be other times, however, when we would like to create a new one from scratch. In this recipe, we will train a new NER model to parse out the names of musicians and their works of art.

Getting ready

We will use the spaCy package to train a new NER model. You do not need any other packages other than `spacy`. The data we are going to use is from `https://github.com/deezer/music-ner-eacl2023`. The data file is preloaded in the data folder (`https://github.com/PacktPublishing/Python-Natural-Language-Processing-Cookbook-Second-Edition/blob/main/data/music_ner.csv`) and you will need to download it from the book's GitHub repository into the `data` directory.

The notebook is located at `https://github.com/PacktPublishing/Python-Natural-Language-Processing-Cookbook-Second-Edition/blob/main/Chapter05/5.5_training_own_spacy_model.ipynb`.

How to do it...

We will define our training data and then use it to train a new model. We will then test the model and save it to disk. The steps are as follows:

1. Run the language utilities file:

   ```
   %run -i "../util/lang_utils.ipynb"
   ```

2. Import other functions and packages:

   ```
   import pandas as pd
   from spacy.cli.train import train
   from spacy.cli.evaluate import evaluate
   from spacy.tokens import DocBin
   from sklearn.model_selection import train_test_split
   ```

3. In this step, we load the data and print it out:

   ```
   music_ner_df = pd.read_csv('../data/music_ner.csv')
   print(music_ner_df)
   ```

The data has five columns: id, start offset, end offset, text, and label. Sentences repeat if there is more than one entity per sentence, as there is one row per named entity. There are 428 entries in the data.

```
        id                                          text  start_offset  \
0    13434  i love radioheads kid a something similar | ki...            7
1    13434  i love radioheads kid a something similar | ki...           61
2    13435              anything similar to i fight dragons           20
3    13436             music similar to ccrs travelin band           17
4    13437               songs similar to blackout by boris           17
..     ...                                          ...          ...
423  14028  songs like good news by mac miller | preferrab...           11
424  14028  songs like good news by mac miller | preferrab...           24
425  14030  something along the lines of either the chain ...           49
426  14030  something along the lines of either the chain ...           29
427  14032         heavy bass x gothic rap like oxygen by bones           29

     end_offset                label
0            17          Artist_known
1            71  Artist_or_WoA_deduced
2            35           WoA_deduced
3            30        Artist_deduced
4            25           WoA_deduced
..          ...                  ...
423          20           WoA_deduced
424          34        Artist_deduced
425          60  Artist_or_WoA_deduced
426          45  Artist_or_WoA_deduced
427          44  Artist_or_WoA_deduced

[428 rows x 5 columns]
```

Figure 5.2 – DataFrame output

4. Here, we remove _deduced from the labels so the labels are now Artist, WoA (work of art), Artist_or_WoA:

```
def change_label(input_label):
    input_label = input_label.replace("_deduced", "")
    return input_label

music_ner_df["label"] = music_ner_df["label"].apply(change_label)
print(music_ner_df)
```

The result will look like this:

```
           id                                              text  start_offset  \
0       13434  i love radioheads kid a something similar | ki...             7
1       13434  i love radioheads kid a something similar | ki...            61
2       13435               anything similar to i fight dragons            20
3       13436               music similar to ccrs travelin band            17
4       13437                  songs similar to blackout by boris           17
..        ...                                               ...           ...
422     14028  songs like good news by mac miller | preferrab...           11
423     14028  songs like good news by mac miller | preferrab...           24
424     14030  something along the lines of either the chain ...           49
425     14030  something along the lines of either the chain ...           29
426     14032             heavy bass x gothic rap like oxygen by bones      29

     end_offset           label
0            17    Artist_known
1            71    Artist_or_WoA
2            35             WoA
3            30          Artist
4            25             WoA
..          ...             ...
422          20             WoA
423          34          Artist
424          60    Artist_or_WoA
425          45    Artist_or_WoA
426          44    Artist_or_WoA

[427 rows x 5 columns]
```

Figure 5.3 – DataFrame output

5. In this step, we create the `DocBin` objects that will store the processed data. `DocBin` objects are required for input data for spaCy models (to learn more about them, see the *Training a spaCy textcat model* recipe in *Chapter 4*):

    ```
    train_db = DocBin()
    test_db = DocBin()
    ```

6. Here, we create a list of unique IDs and split it into training and test data. The reason we would like to get the unique IDs is because sentences repeat through the dataset. There are 227 unique IDs (or sentences), and there are 170 sentences in the training data and 57 sentences in the test data:

    ```
    # Get a unique list of unique ids
    ids = list(set(music_ner_df["id"].values))
    print(len(ids))
    # Split ids into training and test
    ```

```
train_ids, test_ids = train_test_split(ids)
print(len(train_ids))
print(len(test_ids))
```

The result will be as follows:

```
227
170
57
```

7. Here, we create and save training and test data in the DocBin objects. We loop through IDs, and for each ID, we get the sentence. We process the sentence using the small model and then have a spaCy Doc object. Then, we loop through the entities in the sentence and add them to the ents attribute of the Doc object. The processed Doc object then goes into one of the DocBin objects:

```
for id in ids:
    entity_rows = music_ner_df.loc[music_ner_df['id'] == id]
    text = entity_rows.head(1)["text"].values[0]
    doc = small_model(text)
    ents = []
    for index, row in entity_rows.iterrows():
        label = row["label"]
        start = row["start_offset"]
        end = row["end_offset"]
        span = doc.char_span(start, end, label=label,
            alignment_mode="contract")
        ents.append(span)
    doc.ents = ents
    if id in train_ids:
        train_db.add(doc)
    else:
        test_db.add(doc)
train_db.to_disk('../data/music_ner_train.spacy')
test_db.to_disk('../data/music_ner_test.spacy')
```

8. In this step, we train the model. We use the spacy_config_ner.cfg configuration file in the data folder. You can create your own customized configuration file at https://spacy.io/usage/training/#quickstart. The output shows the loss, accuracy, precision, recall, F1 score, and other metrics for every epoch. Finally, it saves the model to the specified directory:

```
train("../data/spacy_config_ner.cfg", output_path="../models/
spacy_music_ner")
```

The output will look like this:

```
ℹ Saving to output directory: ../models/spacy_music_ner
ℹ Using CPU

========================= Initializing pipeline =========================
✓ Initialized pipeline

========================= Training pipeline =========================
ℹ Pipeline: ['tok2vec', 'tagger', 'parser', 'ner']
ℹ Initial learn rate: 0.001
E    #     LOSS TOK2VEC  LOSS TAGGER  LOSS PARSER  LOSS NER  TAG_ACC  DEP_UAS  DEP_LAS  SENTS_F  ENTS_F  ENTS_P  ENTS_R  SCORE
---  ----  ------------  -----------  -----------  --------  -------  -------  -------  -------  ------  ------  ------  ------
  0     0          0.00        85.39       265.35     63.00    35.26    22.92    14.84     3.66    0.00    0.00    0.00    0.18
  5   200        747.41      2873.85     10642.29   3960.21    77.27    69.25    59.52    55.17   37.07   36.19   38.00    0.60
 12   400        700.66       254.46      2183.84    370.70    77.92    72.38    63.15    82.26   44.55   44.12   45.00    0.63
 21   600        914.24       163.11      1288.12    407.55    79.24    69.08    61.67    71.21   39.80   40.62   39.00    0.62
 32   800       1066.64       128.50      1143.84    421.63    78.09    72.22    64.30    79.03   40.41   41.94   39.00    0.62
 45  1000       1464.56       117.67       734.04   1159.56    79.90    69.41    62.32    76.80   43.39   46.07   41.00    0.63
 61  1200        941.88        90.95       917.36     60.92    80.40    69.41    62.32    72.31   42.93   45.05   41.00    0.63
 80  1400       1385.22        87.17       843.28     98.64    79.90    70.90    62.82    77.17   44.92   48.28   42.00    0.64
104  1600        953.55        78.03       586.17    153.99    80.07    71.56    64.63    74.60   45.23   45.45   45.00    0.64
133  1800        970.64        65.13       730.98     63.91    80.40    71.06    62.82    72.31   43.75   45.65   42.00    0.64
168  2000       4200.54        70.03       697.10   1060.07    80.56    69.91    61.34    74.42   40.61   41.24   40.00    0.62
211  2200       4038.66        84.34       721.41   1291.99    80.40    70.57    63.97    76.19   48.08   46.30   50.00    0.65
261  2400        372.10        52.12       670.31      1.77    80.89    72.55    65.46    84.55   47.67   49.46   46.00    0.66
311  2600        546.39        37.73       670.02     20.32    81.22    71.23    64.14    69.17   45.10   44.23   46.00    0.65
361  2800        824.87        32.01       666.86     10.36    81.22    70.73    63.48    81.60   47.18   48.42   46.00    0.65
411  3000       1248.57        29.03       571.53     90.03    80.72    68.76    63.15    68.18   48.98   50.00   48.00    0.65
461  3200        744.53        24.98       604.77     20.52    81.05    69.08    61.67    80.65   51.28   52.63   50.00    0.66
511  3400      10586.64        27.44       607.36   1063.01    79.74    71.89    64.14    70.07   53.06   54.17   52.00    0.67
561  3600       2715.90        29.15       736.47     95.58    80.89    73.37    64.96    73.02   47.92   50.00   46.00    0.66
611  3800       1497.09        24.71       694.38     24.16    79.74    69.58    62.82    75.00   47.47   47.96   47.00    0.64
661  4000       1339.60        22.34       618.29     45.45    81.71    71.23    63.97    79.03   43.06   41.28   45.00    0.64
711  4200       1100.50        20.76       619.86     21.44    80.07    71.72    64.30    70.68   47.67   49.46   46.00    0.65
761  4400       1548.19        19.02       649.54     27.96    80.23    69.41    63.31    80.95   43.75   45.65   42.00    0.63
811  4600        792.71        17.04       533.41     45.45    80.23    71.89    65.46    76.80   47.00   47.00   47.00    0.65
861  4800        785.91        16.64       558.07     10.96    80.07    72.38    64.63    75.97   47.37   50.00   45.00    0.65
911  5000       2307.66        17.14       556.16     86.73    80.07    70.90    63.64    78.74   43.08   44.21   42.00    0.64
✓ Saved pipeline to output directory
../models/spacy_music_ner/model-last
```

Figure 5.4 – Model training output

9. In this step, we load the trained model and use it on data not seen during training. We get an ID from the test set, get all the rows from the test data with that ID, and load the sentence. We then print out the sentence and the annotated entities. Then, we process the sentence using our model (in exactly the same way as other pretrained spaCy models) and print out the entities it parsed:

```
nlp = spacy.load("../models/spacy_music_ner/model-last")
first_test_id = test_ids[0]
test_rows = music_ner_df.loc[music_ner_df['id']
    == first_test_id]
input_text = entity_rows.head(1)["text"].values[0]
print(input_text)
print("Gold entities:")
for index, row in entity_rows.iterrows():
    label = row["label"]
    start = row["start_offset"]
    end = row["end_offset"]
```

```
        span = doc.char_span(start, end, label=label,
            alignment_mode="contract")
        print(span)
doc = nlp(input_text)
print("Predicted entities: ")
for entity in doc.ents:
    print(entity)
```

We see that the resulting entities are quite good (output results may vary):

```
songs with themes of being unable to settle | ex hoziers someone
new elle kings exes and ohs
Gold entities:
hoziers
someone new
elle kings
exes and ohs
Predicted entities:
hoziers
someone new
elle kings
exes and
```

10. Here, we evaluate the model using spaCy's `evaluate` function. We see that the `WoA` and `Artist` tags have metrics that are low but in the double digits, while the `Artist_or_WoA` tag has an F1 score of about 10%. This is due to the fact that it has much less data than the other two tags. Overall, the performance of the model according to the statistics is not very good, and that is because we have a very small amount of data overall:

```
evaluate('../models/spacy_music_ner/model-last', '../data/music_
ner_tes t.spacy')
```

The statistics might vary, but here is the output I got (output condensed):

```
{'token_acc': 1.0,
 'token_p': 1.0,
 'token_r': 1.0,
 'token_f': 1.0,
 'tag_acc': 0.800658978583196,
 ...
 'ents_p': 0.4421052631578947,
 'ents_r': 0.42,
 'ents_f': 0.4307692307692308,
 'ents_per_type': {'WoA': {'p': 0.4358974358974359,
   'r': 0.425,
   'f': 0.43037974683544306},
  'Artist_or_WoA': {'p': 0.1,
```

```
     'r': 0.09090909090909091,
     'f': 0.09523809523809525},
   'Artist': {'p': 0.5217391304347826,
     'r': 0.4897959183673469,
     'f': 0.5052631578947369}},
 'speed': 3835.591242612551}
```

See also

The spaCy NER model is a neural network model. You can learn more about its architecture from the spaCy documentation: `https://spacy.io/models#architecture`.

Fine-tuning BERT for NER

In this recipe, we will fine-tune the pretrained BERT model for the NER task. The difference between training a model from scratch and fine-tuning it is as follows. Fine-tuning an NLP model, such as BERT, involves taking a pretrained model and modifying it for your specific task, such as NER in this case. The pretrained model already has lots of knowledge stored in it and the results are likely to be better than when training a model from scratch.

We will use similar data as in the previous recipe, creating a model that can tag entities as `Artist` or `WoA`. The data comes from the same dataset but it is labeled using the IOB format, which is required for the `transformers` packages we are going to use. We also only use the `Artist` and `WoA` tags, removing the `Artist_or_WoA` tag, since there is not enough data for that tag.

For this recipe, we will use the Hugging Face `Trainer` class, although it is also possible to train Hugging Face models using PyTorch or Tensorflow. See more at `https://huggingface.co/docs/transformers/training`.

Getting ready

We will use the `transformers` package from Hugging Face. It is preloaded in the Poetry environment. You can also install the package from the `requirements.txt` file.

The notebook is located at `https://github.com/PacktPublishing/Python-Natural-Language-Processing-Cookbook-Second-Edition/blob/main/Chapter05/5.6_fine_tune_bert.ipynb`.

How to do it...

We will load and preprocess the data, train the model, and evaluate it, and then we will use it on unseen data. Your steps should be formatted like so:

1. Run the language utilities notebook:

```
%run -i "../util/lang_utils.ipynb"
```

2. Import other packages and functions:

```
from datasets import (
    load_dataset, Dataset, Features, Value,
    ClassLabel, Sequence, DatasetDict)
import pandas as pd
from transformers import AutoTokenizer, AutoModel
from transformers import DataCollatorForTokenClassification
from transformers import (
    AutoModelForTokenClassification,
    TrainingArguments, Trainer)
import numpy as np
from sklearn.model_selection import train_test_split
from evaluate import load
```

3. In this step, we load the music NER dataset using the pandas `read_csv` function. We then define a function that takes a label, splits it on the underscore, and removes the last part (`_deduced`). We then apply this function to the `label` column. We also substitute the `|` character in case it could interfere with our code:

```
music_ner_df = pd.read_csv('../data/music_ner.csv')
def change_label(input_label):
    input_label = input_label.replace("_deduced", "")
    return input_label
music_ner_df["label"] = music_ner_df["label"].apply(
    change_label)
music_ner_df["text"] = music_ner_df["text"].apply(
    lambda x: x.replace("|", ","))
print(music_ner_df)
```

The output will look similar to this:

```
         id                                              text  start_offset  \
0     13434  i love radioheads kid a something similar , ki...             7
1     13434  i love radioheads kid a something similar , ki...            61
2     13435                 anything similar to i fight dragons            20
3     13436                 music similar to ccrs travelin band            17
4     13437                  songs similar to blackout by boris            17
..      ...                                                ...           ...
422   14028  songs like good news by mac miller , preferrab...            11
423   14028  songs like good news by mac miller , preferrab...            24
424   14030  something along the lines of either the chain ...            49
425   14030  something along the lines of either the chain ...            29
426   14032          heavy bass x gothic rap like oxygen by bones            29

     end_offset         label
0            17        Artist
1            71  Artist_or_WoA
2            35           WoA
3            30        Artist
4            25           WoA
..          ...           ...
422          20           WoA
423          34        Artist
424          60  Artist_or_WoA
425          45  Artist_or_WoA
426          44  Artist_or_WoA

[427 rows x 5 columns]
```

Figure 5.5 – Dataset DataFrame

4. Here, we start our data preprocessing. We get the list of unique IDs from the `id` column. We loop through this list and get the sentence that corresponds to the given ID. We then process the text using the spaCy small model and add the entities from the DataFrame into the `Doc` object. We then store each of these sentences in a dictionary in which the keys are the sentence text strings and the values are the `Doc` objects:

```
ids = list(set(music_ner_df["id"].values))
docs = {}
for id in ids:
    entity_rows = music_ner_df.loc[music_ner_df['id'] == id]
    text = entity_rows.head(1)["text"].values[0]
    doc = small_model(text)
    ents = []
    for index, row in entity_rows.iterrows():
        label = row["label"]
        start = row["start_offset"]
```

```
        end = row["end_offset"]
        span = doc.char_span(start, end, label=label,
            alignment_mode="contract")
        ents.append(span)
    doc.ents = ents
    docs[doc.text] = doc
```

5. Now, we load the data in IOB format. This format is required to fine-tune BERT, as opposed to the format used by spaCy. For that, we load a separate data file, `../data/music_ner_bio.bio`. We create a dictionary of tag mappings and initialize empty lists for tokens, NER tags, and spans. We then loop through the sentence data we read from the data file. For each sentence, each line is a pair of a word and its label. We append the words to the `words` list and the numbers corresponding to the labels to the `tags` list. We also get the spans from the dictionary of `Doc` objects we created in the previous step and append those to the `spans` list:

```
data_file = "../data/music_ner_bio.bio"
tag_mapping = {"O": 0, "B-Artist": 1, "I-Artist": 2,
    "B-WoA": 3, "I-WoA": 4}
with open(data_file) as f:
    data = f.read()
tokens = []
ner_tags = []
spans = []
sentences = data.split("\n\n")
for sentence in sentences:
    words = []
    tags = []
    this_sentence_spans = []
    word_tag_pairs = sentence.split("\n")
    for pair in word_tag_pairs:
        (word, tag) = pair.split("\t")
        words.append(word)
        tags.append(tag_mapping[tag])
    sentence_text = " ".join(words)
    try:
        doc = docs[sentence_text]
    except:
        pass
    ent_dict = {}
    for ent in doc.ents:
        this_sentence_spans.append(f"{ent.label_}: {ent.text}")
    tokens.append(words)
    ner_tags.append(tags)
    spans.append(this_sentence_spans)
```

6. Here, we split the data into training and testing. For that, we split the indices of the spans list. Then, we create separate tokens, NER tags, and spans lists for training and test data:

```
indices = range(0, len(spans))
train, test = train_test_split(indices, test_size=0.1)
train_tokens = []
test_tokens = []
train_ner_tags = []
test_ner_tags = []
train_spans = []
test_spans = []
for i, (token, ner_tag, span) in enumerate(
    zip(tokens, ner_tags, spans)
):
    if i in train:
        train_tokens.append(token)
        train_ner_tags.append(ner_tag)
        train_spans.append(span)
    else:
        test_tokens.append(token)
        test_ner_tags.append(ner_tag)
        test_spans.append(span)

print(len(train_spans))
print(len(test_spans))
```

The output will be as follows:

```
539
60
```

7. In this step, we create new DataFrames from the training and test lists we compiled in *step 6*. We then join the contents of the tokens column with spaces to get a sentence string instead of a list of words. We then drop empty data using the dropna() function and print the contents of the test DataFrame:

```
training_df = pd.DataFrame({"tokens":train_tokens,
    "ner_tags": train_ner_tags, "spans": train_spans})
test_df = pd.DataFrame({"tokens": test_tokens,
    "ner_tags": test_ner_tags, "spans": test_spans})
training_df["text"] = training_df["tokens"].apply(
    lambda x: " ".join(x))
test_df["text"] = test_df["tokens"].apply(lambda x: " ".join(x))
training_df.dropna()
test_df.dropna()
print(test_df)
```

The result will look like this:

```
                                                    tokens  \
    0    [i, love, radioheads, kid, a, something, simil...
    1    [bluesy, songs, kinda, like, evil, woman, by, ...
    ...
    58   [looking, for, like, electronic, music, with, ...
    59   [looking, for, pop, songs, about, the, end, of...


                                                   ner_tags  \
    0           [0, 0, 1, 3, 4, 0, 0, 0, 0, 0, 0, 0, 1, 2, 0]
    1                             [0, 0, 0, 0, 3, 4, 0, 1, 2]
    ...
    58          [0, 0, 0, 0, 0, 0, 0, 0, 0, 0, 0, 0, 0, 0, 0]
    59                      [0, 0, 0, 0, 0, 0, 0, 0, 0, 0]


                                                     spans  \
    0      [Artist: radioheads, Artist_or_WoA: aphex twin]
    1            [WoA: evil woman, Artist: black sabbath]
    ...
    58   [WoA: the piper at the gates of dawn, Artist: ...
    59   [WoA: the piper at the gates of dawn, Artist: ...


                                                      text
    0    i love radioheads kid a something similar , ki...
    1    bluesy songs kinda like evil woman by black sa...
    ...
    58   looking for like electronic music with a depre...
    59     looking for pop songs about the end of the world
```

8. Here, we load the pretrained model and tokenizer and initialize the Dataset objects. The Features object describes the data and its properties. We create one training and one test Dataset object. We use the Features object we created, and the DataFrames we initialized in the previous step. We then add these newly created Dataset objects to DatasetDict, with one entry for the training dataset and one for the test data. We then print out the resulting object:

```
tokenizer = AutoTokenizer.from_pretrained('bert-base-uncased')
features = Features(
    {'tokens': Sequence(feature=Value(dtype='string',
            id=None),
        length=-1, id=None),
            'ner_tags': Sequence(feature=ClassLabel(
                names=['O', 'B-Artist', 'I-Artist',
                'B-WoA', 'I-WoA'], id=None),
                length=-1, id=None),
```

```
                    'spans': Sequence(
                        feature=Value(dtype='string',id=None),
                        length=-1, id=None),
                    'text': Value(dtype='string', id=None)
                              })
    training_dataset = Dataset.from_pandas(
        training_df, features=features)
    test_dataset = Dataset.from_pandas(test_df, features=features)
    dataset = DatasetDict({"train":training_dataset,
        "test":test_dataset})
    print(dataset["train"].features)
    label_names = \
        dataset["train"].features["ner_tags"].feature.names
    print(dataset)
```

The result will be as follows:

```
{'tokens': Sequence(feature=Value(dtype='string',
id=None), length=-1, id=None), 'ner_tags':
Sequence(feature=ClassLabel(names=['O', 'B-Artist', 'I-Artist',
'B-WoA', 'I-WoA'], id=None), length=-1, id=None), 'spans':
Sequence(feature=Value(dtype='string', id=None), length=-1,
id=None), 'text': Value(dtype='string', id=None)}
DatasetDict({
    train: Dataset({
        features: ['tokens', 'ner_tags', 'spans', 'text'],
        num_rows: 539
    })
    test: Dataset({
        features: ['tokens', 'ner_tags', 'spans', 'text'],
        num_rows: 60
    })
})
```

9. In this step, we create the `tokenize_adjust_labels` function that will assign the correct labels to word parts. We define the `tokenize_adjust_labels` function. The BERT tokenizer splits some words into components, and we need to make sure that the same label is assigned to each word part. The function first tokenizes all the text samples using the preloaded tokenizer. It then loops through the input IDs of the tokenized samples and adjusts the labels according to the word parts:

```
def tokenize_adjust_labels(all_samples_per_split):
    tokenized_samples = tokenizer.batch_encode_plus(
    all_samples_per_split["text"])
    total_adjusted_labels = []
    for k in range(0, len(tokenized_samples["input_ids"])):
        prev_wid = -1
```

```
            word_ids_list = tokenized_samples.word_ids(
                batch_index=k)
            existing_label_ids = all_samples_per_split[
                "ner_tags"][k]
            i = -1
            adjusted_label_ids = []
            for wid in word_ids_list:
                if (wid is None):
                    adjusted_label_ids.append(-100)
                elif (wid != prev_wid):
                    i = i + 1
                    adjusted_label_ids.append(existing_label_ids[i])
                    prev_wid = wid
                else:
                    label_name =label_names[existing_label_ids[i]]
                    adjusted_label_ids.append(existing_label_ids[i])
            total_adjusted_labels.append(adjusted_label_ids)
        tokenized_samples["labels"] = total_adjusted_labels
        return tokenized_samples
```

10. Use the previous function on the dataset:

```
tokenized_dataset = dataset.map(tokenize_adjust_labels,
    batched=True)
```

11. Here, we initialize the data collator object. Data collators simplify the handling of data for training, for example, padding and truncating the input for all inputs to be of the same length:

```
data_collator = DataCollatorForTokenClassification(tokenizer)
```

12. Now, we create the `compute_metrics` function, which calculates evaluation metrics, including precision, recall, F1 score, and accuracy. In the function, we delete all the tokens that have the label -100, which are the special tokens. This function uses the `seqeval` evaluation method commonly used to evaluate NER tasks:

```
metric = load("seqeval")
def compute_metrics(data):
    predictions, labels = data
    predictions = np.argmax(predictions, axis=2)

    data = zip(predictions, labels)
    data = [
        [(p, l) for (p, l) in zip(prediction, label)
            if l != -100]
        for prediction, label in data
    ]
```

```
true_predictions = [
    [label_names[p] for (p, l) in data_point]
    for data_point in data
]
true_labels = [
    [label_names[l] for (p, l) in data_point]
    for data_point in data
]

results = metric.compute(predictions=true_predictions,
    references=true_labels)
flat_results = {
    "overall_precision": results["overall_precision"],
    "overall_recall": results["overall_recall"],
    "overall_f1": results["overall_f1"],
    "overall_accuracy": results["overall_accuracy"],
}
for k in results.keys():
  if (k not in flat_results.keys()):
    flat_results[k + "_f1"] = results[k]["f1"]

return flat_results
```

13. Here, we load the pretrained BERT model (the uncased version since our input is in lowercase). We then specify the training arguments by initializing the `TrainingArguments` object. This object contains the model hyperparameters. We then initialize the `Trainer` object by providing the training arguments, dataset, tokenizer, data collator, and `metrics` function. We then start the training process:

```
model = AutoModelForTokenClassification.from_pretrained(
    'bert-base-uncased', num_labels=len(label_names))
training_args = TrainingArguments(
    output_dir="./fine_tune_bert_output",
    evaluation_strategy="steps",
    learning_rate=2e-5,
    per_device_train_batch_size=16,
    per_device_eval_batch_size=16,
    num_train_epochs=7,
    weight_decay=0.01,
    logging_steps = 1000,
    run_name = "ep_10_tokenized_11",
    save_strategy='no'
)
```

```
trainer = Trainer(
    model=model,
    args=training_args,
    train_dataset=tokenized_dataset["train"],
    eval_dataset=tokenized_dataset["test"],
    data_collator=data_collator,
    tokenizer=tokenizer,
    compute_metrics=compute_metrics
)
trainer.train()
```

The output will include different information, including the following:

```
TrainOutput(global_step=238, training_loss=0.25769581514246326,
metrics={'train_runtime': 25.8951, 'train_samples_per_second':
145.703, 'train_steps_per_second': 9.191, 'total_flos':
49438483110900.0, 'train_loss': 0.25769581514246326, 'epoch':
7.0})
```

14. In this step, we evaluate the fine-tuned model:

```
trainer.evaluate()
```

For the Artist label, it achieves an F1 score of 76%, and for the WoA label, it achieves an F1 score of 52%:

```
{'eval_loss': 0.28670933842658997,
 'eval_overall_precision': 0.6470588235294118,
 'eval_overall_recall': 0.7096774193548387,
 'eval_overall_f1': 0.6769230769230768,
 'eval_overall_accuracy': 0.9153605015673981,
 'eval_Artist_f1': 0.761904761904762,
 'eval_WoA_f1': 0.5217391304347826,
 'eval_runtime': 0.3239,
 'eval_samples_per_second': 185.262,
 'eval_steps_per_second': 12.351,
 'epoch': 7.0}
```

15. Save the model:

```
trainer.save_model("../models/bert_fine_tuned")
```

16. Now, load the trained model:

```
model = AutoModelForTokenClassification.from_pretrained("../
models/bert_fine_tuned")
tokenizer = AutoTokenizer.from_pretrained(
    "../models/bert_fine_tuned")
```

17. Here, we test the fine-tuned model on an unseen text. We initialize the `text` variable. We then load the `pipeline` package to create a pipeline we will use. A text-processing pipeline takes the text to its final output value processed by the model. This particular pipeline specifies the task as `token-classification`, which fine-tuned model to use, the corresponding tokenizer, and the aggregation strategy. The aggregation strategy parameter specifies how to combine the results of several models when several models are used. We then run the pipeline on the text:

```
text = "music similar to morphine robocobra quartet | featuring
elements like saxophone prominent bass"
from transformers import pipeline
pipe = pipeline(task="token-classification",
    model=model.to("cpu"), tokenizer=tokenizer,
    aggregation_strategy="simple")
pipe(text)
# tag_mapping = {"O": 0, "B-Artist": 1, "I-Artist": 2, "B-WoA":
3, "I-WoA": 4}
```

The output will vary. The sample output identifies the music artist, *Morphine Robocobra Quartet*:

```
[{'entity_group': 'LABEL_0',
  'score': 0.9991929,
  'word': 'music similar to',
  'start': 0,
  'end': 16},
 {'entity_group': 'LABEL_1',
  'score': 0.8970744,
  'word': 'morphine robocobra',
  'start': 17,
  'end': 35},
 {'entity_group': 'LABEL_2',
  'score': 0.5060059,
  'word': 'quartet',
  'start': 36,
  'end': 43},
 {'entity_group': 'LABEL_0',
  'score': 0.9988042,
  'word': '| featuring elements like saxophone prominent bass',
  'start': 44,
  'end': 94}]
```

We can see that the labels assigned by the model are correct.

6
Topic Modeling

In this chapter, we will cover **topic modeling**, or the classification of topics present in a corpus of text. Topic modeling is a very useful technique that can give us an idea about which topics appear in a document set. For example, topic modeling is used for trend discovery on social media. Also, in many cases, it is useful to do topic modeling as part of the preliminary data analysis of a dataset to understand which topics appear in it.

There are many different algorithms available to do this. All of them try to find similarities between different texts and put them into several clusters. These different clusters indicate different topics.

You will learn how to create and use topic models via various techniques with the BBC news dataset in this chapter. This dataset has news that falls within the following topics: politics, sport, business, tech, and entertainment. Thus, we know that in each case, we need to have five topic clusters. This is not going to be the case in real-life scenarios, and you will need to estimate the number of topic clusters. A good reference on how to do this is *The Hundred-Page Machine Learning Book* by Andriy Burkov (p. 112).

This chapter contains the following recipes:

- LDA topic modeling with `gensim`
- Community detection clustering with SBERT
- K-Means topic modeling with BERT
- Topic modeling using BERTopic
- Using contextualized topic models

Technical requirements

In this chapter, we will work with the same BBC dataset that we worked with in *Chapter 4*. The dataset is located in the book GitHub repository:

```
https://github.com/PacktPublishing/Python-Natural-Language-Processing-
Cookbook-Second-Edition/blob/main/data/bbc-text.csv
```

It is also available through Hugging Face:

```
https://huggingface.co/datasets/SetFit/bbc-news
```

> **Note**
>
> This dataset is used in this book with permission from the researchers. The original paper associated with this dataset is as follows:
>
> Derek Greene and Pádraig Cunningham. "Practical Solutions to the Problem of Diagonal Dominance in Kernel Document Clustering", in Proc. 23rd International Conference on Machine Learning (ICML'06), 2006.
>
> All rights, including copyright, in the text content of the original articles are owned by the BBC.

Please make sure to download all the Python notebooks in the `util` folder on GitHub into the `util` folder on your computer. The directory structure on your computer should mirror the setup in the GitHub repository. We will be accessing files in this directory in several recipes in this chapter.

LDA topic modeling with gensim

Latent Dirichlet Allocation (**LDA**) is one of the oldest algorithms for topic modeling. It is a statistical generative model that calculates the probabilities of different words. In general, LDA is a good choice of model for longer texts.

We will use one of the main topic modeling algorithms, LDA, to create a topic model for the BBC news texts. We know that the BBC news dataset has five topics: tech, politics, business, entertainment, and sport. Thus, we will use five as the expected number of clusters.

Getting ready

We will be using the `gensim` package, which is part of the poetry environment. You can also install the `requirements.txt` file to get the package.

The dataset is located at `https://github.com/PacktPublishing/Python-Natural-Language-Processing-Cookbook-Second-Edition/blob/main/data/bbc-text.csv` and should be downloaded to the `data` folder.

The notebook is located at `https://github.com/PacktPublishing/Python-Natural-Language-Processing-Cookbook-Second-Edition/blob/main/Chapter06/6.1_topic_modeling_gensim.ipynb`.

How to do it...

An LDA model requires the data to be clean. This means that stopwords and other unnecessary tokens need to be removed from the text. This includes digits and punctuation. If this step is skipped, topics that center around stopwords, digits, or punctuation might appear.

We will load the data, clean it, and preprocess it using the `simple_preprocess` function, which is available through the `gensim` package. Then we will create the LDA model. Any topic model requires the engineer to estimate the number of topics in advance. We will use five, as we know that there are five topics present in the data. For more information on how to estimate the number of topics, please see the introductory section of this chapter.

The steps are as follows:

1. Do the necessary imports:

```
import pandas as pd
import nltk
import re
from nltk.corpus import stopwords
from gensim.utils import simple_preprocess
from gensim.models.ldamodel import LdaModel
import gensim.corpora as corpora
from pprint import pprint
from gensim.corpora import MmCorpus
```

2. Load the stopwords and the BBC news data, then print the resulting dataframe. Here, we use the standard stopword list from NLTK. As we saw in *Chapter 4*, in the *Clustering sentences using K-Means: unsupervised text classification* recipe, the `said` word is also considered a stopword in this dataset, so we must manually add it to the list.

```
stop_words = stopwords.words('english')
stop_words.append("said")
bbc_df = pd.read_csv("../data/bbc-text.csv")
print(bbc_df)
```

The result will look similar to this:

```
         category                                                    text
0            tech    tv future in the hands of viewers with home th...
1        business    worldcom boss  left books alone  former worldc...
2           sport    tigers wary of farrell  gamble  leicester say ...
3           sport    yeading face newcastle in fa cup premiership s...
4   entertainment    ocean s twelve raids box office ocean s twelve...
...           ...                                                    ...
2220     business    cars pull down us retail figures us retail sal...
2221     politics    kilroy unveils immigration policy ex-chatshow ...
2222 entertainment    rem announce new glasgow concert us band rem h...
2223     politics    how political squabbles snowball it s become c...
2224        sport    souness delight at euro progress boss graeme s...

[2225 rows x 2 columns]
```

Figure 6.1 – The BBC news dataframe output

3. In this step, we will create the `clean_text` function. This function removes extra whitespace in the first line and digits in the second line from text. It then uses the `simple_preprocess` function from gensim. The `simple_preprocess` function splits the text into a list of tokens, lowercases them, and removes tokens that are too long or too short. We then remove stopwords from the list:

```
def clean_text(input_string):
    input_string = re.sub(r'[^\w\s]', ' ', input_string)
    input_string = re.sub(r'\d', '', input_string)
    input_list = simple_preprocess(input_string)
    input_list = [word for word in input_list if word not in
        stop_words]
    return input_list
```

4. We will now apply the function to the data. The text column now contains a list of words that are in lowercase and no stopwords:

```
bbc_df['text'] = bbc_df['text'].apply(lambda x: clean_text(x))
print(bbc_df)
```

The result will look similar to this:

```
       category                                              text
0          tech  [tv, future, hands, viewers, home, theatre, sy...
1      business  [worldcom, boss, left, books, alone, former, w...
2         sport  [tigers, wary, farrell, gamble, leicester, say...
3         sport  [yeading, face, newcastle, fa, cup, premiershi...
4 entertainment  [ocean, twelve, raids, box, office, ocean, twe...
...             ...                                              ...
2220   business  [cars, pull, us, retail, figures, us, retail, ...
2221   politics  [kilroy, unveils, immigration, policy, ex, cha...
2222 entertainment  [rem, announce, new, glasgow, concert, us, ban...
2223   politics  [political, squabbles, snowball, become, commo...
2224      sport  [souness, delight, euro, progress, boss, graem...

[2225 rows x 2 columns]
```

Figure 6.2 – The processed BBC news output

Here, we will use the `gensim.corpora.Dictionary` class to create a mapping from a word to its integer ID. This is necessary to then create a bag-of-words representation of the text. Then, using this mapping, we create the corpus as a bag of words. To learn more about the bag-of-words concept, please see *Chapter 3, Putting Documents into a Bag of Words*. In this recipe, instead of using `sklean`'s `CountVectorizer` class, we will use the classes provided by the `gensim` package:

```
texts = bbc_df['text'].values
id_dict = corpora.Dictionary(texts)
corpus = [id_dict.doc2bow(text) for text in texts]
```

5. In this step, we will initialize and train the LDA model. We will pass in the preprocessed and vectorized data (`corpus`), the word-to-ID mapping (`id_dict`), the number of topics, which we initialized to five, the chunk size, and the number of passes. The chunk size determines the number of documents used in each training chunk, and the number of passes specifies the number of passes through the corpus during training. You can experiment with these hyperparameters to see whether they improve the model. The parameters used here, 100 documents per chunk and 20 passes, were chosen experimentally to produce a good model:

```
num_topics = 5
lda_model = LdaModel(corpus=corpus,
                     id2word=id_dict,
                     num_topics=num_topics,
                     chunksize=100,
                     passes=20)
```

Here, we will print the resulting topics using the pretty print (`pprint`) function that arranges them nicely. The results will vary each time you train the model. For this training pass, topic 0 is tech, topic 1 is sport, topic 2 is business, topic 3 is entertainment, and topic 4 is politics:

```
pprint(lda_model.print_topics())
```

The results will vary. Our output looks like this:

```
[(0,
 '0.010*"people" + 0.006*"mobile" + 0.005*"one" + 0.005*"new" + '
 '0.005*"technology" + 0.004*"also" + 0.004*"use" + 0.004*"net" + '
 '0.004*"many" + 0.004*"digital"'),
 (1,
 '0.007*"game" + 0.006*"first" + 0.005*"time" + 0.005*"year" + '
 '0.005*"england" + 0.005*"world" + 0.005*"win" + 0.005*"last" + '
 '0.005*"players" + 0.005*"one"'),
 (2,
 '0.013*"bn" + 0.013*"us" + 0.011*"year" + 0.009*"sales" + 0.006*"market" + '
 '0.006*"company" + 0.005*"growth" + 0.005*"bank" + 0.005*"firm" + '
 '0.005*"also"'),
 (3,
 '0.009*"film" + 0.009*"blair" + 0.008*"party" + 0.008*"best" + 0.006*"also" '
 '+ 0.006*"one" + 0.005*"year" + 0.005*"show" + 0.004*"new" + 0.004*"us"'),
 (4,
 '0.020*"mr" + 0.013*"would" + 0.008*"government" + 0.007*"people" + '
 '0.006*"labour" + 0.005*"could" + 0.005*"election" + 0.005*"minister" + '
 '0.004*"told" + 0.004*"also"')]
```

Figure 6.3 – Our LDA output

There's more...

Now let's save the model and apply it to novel input:

1. Define the `save_model` function. To save the model, we need the model itself, the path where we want to save it, the vectorizer (`id_dict`), the path where we want to save the vectorizer, the corpus, and the path where the corpus will be saved. The function will save these three components to their corresponding paths:

```
def save_model(lda, lda_path, id_dict, dict_path,
    corpus, corpus_path):
    lda.save(lda_path)
    id_dict.save(dict_path)
    MmCorpus.serialize(corpus_path, corpus)
```

2. Save the model, the word-to-ID dictionary, and the vectorized corpus by using the function we just defined:

```
model_path = "../models/bbc_gensim/lda.model"
dict_path = "../models/bbc_gensim/id2word.dict"
```

```
corpus_path = "../models/bbc_gensim/corpus.mm"
save_model(lda_model, model_path, id_dict, dict_path,
    corpus, corpus_path)
```

3. Load the saved model:

```
lda_model = LdaModel.load(model_path)
id_dict = corpora.Dictionary.load(dict_path)
```

4. Define a new example for testing. This example is on the topic of sports:

```
new_example = """Manchester United players slumped to the turf
at full-time in Germany on Tuesday in acknowledgement of what
their
latest pedestrian first-half display had cost them. The 3-2 loss
at
RB Leipzig means United will not be one of the 16 teams in the
draw
for the knockout stages of the Champions League. And this is not
the
only price for failure. The damage will be felt in the accounts,
in
the dealings they have with current and potentially future
players
and in the faith the fans have placed in manager Ole Gunnar
Solskjaer.
With Paul Pogba's agent angling for a move for his client and
ex-United
defender Phil Neville speaking of a "witchhunt" against his
former team-mate
Solskjaer, BBC Sport looks at the ramifications and reaction to
a big loss for United."""
```

5. In this step, we will preprocess the text using the same `clean_text` function, then convert it into a bag-of-words vector and run it through the model. The prediction is a list of tuples, where the first element in each tuple is the number of the topic and the second element is the probability that this text belongs to this particular topic. In this example, we can see that topic 1 is the most probable, which is sport, and is the correct identification:

```
input_list = clean_text(new_example)
bow = id_dict.doc2bow(input_list)
topics = lda_model[bow]
print(topics)
```

The results will vary. Our results look like this:

```
[(1, 0.7338447), (2, 0.15261793), (4, 0.1073401)]
```

Community detection clustering with SBERT

In this recipe, we will use the community detection algorithm included with the **SentenceTransformers** (**SBERT**) package. SBERT will allow us to easily encode sentences using the BERT model. See the *Using BERT and OpenAI embeddings instead of word embeddings* recipe in *Chapter 3* for a more detailed explanation of how to use the sentence transformers.

This algorithm is frequently used to find **communities** in social media but can also be used for topic modeling. The advantage of this algorithm is that it is very fast. It works best on shorter texts, such as texts found on social media. It also only discovers the main topics in the document dataset, as opposed to LDA, which clusters all available text. One use of the community detection algorithm is finding duplicate posts on social media.

Getting ready

We will use the SBERT package in this recipe. It is included in the poetry environment. You can also install it together with other packages by installing the `requirements.txt` file.

The notebook is located at `https://github.com/PacktPublishing/Python-Natural-Language-Processing-Cookbook-Second-Edition/blob/main/Chapter06/6.2_community_detection.ipynb`.

How to do it...

We will transform the text using the BERT sentence transformers model and then use the community detection clustering algorithm on the resulting embeddings.

1. Do the necessary imports. Here, you might need to download the stopwords corpus from NLTK. Please see the *Removing stopwords* recipe in *Chapter 1* for detailed instructions on how to do this.

    ```
    import pandas as pd
    import nltk
    import re
    from nltk.corpus import stopwords
    from sentence_transformers import SentenceTransformer, util
    ```

2. Run the `language utilities` file:

    ```
    %run -i "../util/lang_utils.ipynb"
    ```

3. Load the BBC data and print it:

    ```
    bbc_df = pd.read_csv("../data/bbc-text.csv")
    print(bbc_df)
    ```

    The result will look like *Figure 6.1*.

4. Load the model and create the embeddings. See the *Using BERT and OpenAI embeddings instead of word embeddings* recipe in *Chapter 3* for more information on sentence embeddings. The community detection algorithm requires the embeddings to be in the form of tensors; hence, we must set `convert_to_tensor` to True:

```
model = SentenceTransformer('all-MiniLM-L6-v2')
embeddings = model.encode(bbc_df["text"], convert_to_
tensor=True)
```

5. In this step, we will create the clusters. We will specify the threshold for similarity to be 0.7 on a scale from 0 to 1. This will make sure that the resulting communities are very similar to each other. The minimum community size is 10; that means that a minimum of 10 news articles are required to form a cluster. If we want larger, more general clusters, we should use a larger number for the minimum community size. A more granular clustering should use a smaller number. Any cluster with fewer members will not appear in the output. The result is a list of lists, where each inner list represents a cluster and lists the row IDs of cluster members in the original dataframe:

```
clusters = util.community_detection(
    embeddings, threshold=0.7, min_community_size=10)
print(clusters)
```

The result will vary and might look like this:

```
[[117, 168, 192, 493, 516, 530, 638, 827, 883, 1082, 1154, 1208,
1257, 1359, 1553, 1594, 1650, 1898, 1938, 2059, 2152], [76,
178, 290, 337, 497, 518, 755, 923, 1057, 1105, 1151, 1172, 1242,
1560, 1810, 1813, 1882, 1942, 1981], [150, 281, 376, 503, 758,
900, 1156, 1405, 1633, 1636, 1645, 1940, 1946, 1971], [389, 399,
565, 791, 1014, 1018, 1259, 1288, 1440, 1588, 1824, 1917, 2024],
[373, 901, 1004, 1037, 1041, 1323, 1499, 1534, 1580, 1621, 1751,
2178], [42, 959, 1063, 1244, 1292, 1304, 1597, 1915, 2081, 2104,
2128], [186, 193, 767, 787, 1171, 1284, 1625, 1651, 1797, 2148],
[134, 388, 682, 1069, 1476, 1680, 2106, 2129, 2186, 2198]]
```

6. Here, we will define a function that prints the most common words by cluster. We will take the clusters created by the community detection algorithm and the original dataframe. For each cluster, we will first select the sentences that represent it and then get the most frequent words by using the `get_most_frequent_words` function, which we defined in *Chapter 4*, in the *Clustering sentences using K-Means: unsupervised text classification* recipe. This function is also located in the `lang_utils` notebook that we ran in the second step:

```
def print_words_by_cluster(clusters, input_df):
    for i, cluster in enumerate(clusters):
        print(f"\nCluster {i+1}, {len(cluster)} elements ")
        sentences = input_df.iloc[cluster]["text"]
        all_text = " ".join(sentences)
        freq_words = get_most_frequent_words(all_text)
        print(freq_words)
```

7. Now, use the function on the model output (truncated). We can see that there are many more specific clusters than just the five topics in the original BBC dataset:

```
Cluster 1, 21 elements
['mr', 'labour', 'brown', 'said', 'blair', 'election',
'minister', 'prime', 'chancellor', 'would', 'party', 'new',
'campaign', 'told', 'government', ...]

Cluster 2, 19 elements
['yukos', 'us', 'said', 'russian', 'oil', 'gazprom', 'court',
'rosneft', 'russia', 'yugansk', 'company', 'bankruptcy',
'auction', 'firm', 'unit', ...]

Cluster 3, 14 elements
['kenteris', 'greek', 'thanou', 'iaaf', 'said', 'athens',
'tests', 'drugs', 'olympics', 'charges', 'also', 'decision',
'test', 'athletics', 'missing', ...]

Cluster 4, 13 elements
['mr', 'tax', 'howard', 'labour', 'would', 'said', 'tory',
'election', 'government', 'taxes', 'blair', 'spending',
'tories', 'party', 'cuts',...]

Cluster 5, 12 elements
['best', 'film', 'aviator', 'director', 'actor', 'foxx',
'swank', 'actress', 'baby', 'million', 'dollar', 'said', 'win',
'eastwood', 'jamie',...]

Cluster 6, 11 elements
['said', 'prices', 'market', 'house', 'uk', 'figures',
'mortgage', 'housing', 'year', 'lending', 'november', 'price',
'december', 'rise', 'rose', ...]

Cluster 7, 10 elements
['lse', 'deutsche', 'boerse', 'bid', 'euronext', 'said',
'exchange', 'london', 'offer', 'stock', 'would', 'also',
'shareholders', 'german', 'market',...]

Cluster 8, 10 elements
['dollar', 'us', 'euro', 'said', 'currency', 'deficit',
'analysts', 'trading', 'yen', 'record', 'exports', 'economic',
'trade', 'markets', 'european',...]
```

K-Means topic modeling with BERT

In this recipe, we will use the K-Means algorithm to do unsupervised topic classification, using the BERT embeddings to encode the data. This recipe shares many commonalities with the *Clustering sentences using K-Means – unsupervised text classification* recipe in *Chapter 4*.

The K-Means algorithm is used to find similar clusters with any kind of data and is an easy way to see trends in the data. It is frequently used while performing preliminary data analysis to quickly check the different types of data that appear in a dataset. We can use it with text data and encode the data using a sentence transformer model.

Getting ready

We will be using the `sklearn.cluster.KMeans` object to do the unsupervised clustering, as well as using HuggingFace `sentence transformers`. Both packages are part of the poetry environment.

The notebook is located at `https://github.com/PacktPublishing/Python-Natural-Language-Processing-Cookbook-Second-Edition/blob/main/Chapter06/6.3-kmeans_with_bert.ipynb`.

How to do it...

In the recipe, we will load the BBC dataset and encode it by using the sentence transformers package. We will then use the K-Means clustering algorithm to create five clusters. After that, we will test the model on the test set to see how well it would perform on unseen data:

1. Do the necessary imports:

```
import re
import string
import pandas as pd
from sklearn.model_selection import train_test_split
from sklearn.cluster import KMeans
from nltk.probability import FreqDist
from nltk.corpus import stopwords
from sentence_transformers import SentenceTransformer
```

2. Run the language utilities file. This will allow us to reuse the `print_most_common_words_by_cluster` function in this recipe:

```
%run -i "../util/lang_utils.ipynb"
```

3. Read and print the data:

```
bbc_df = pd.read_csv("../data/bbc-text.csv")
print(bbc_df)
```

The result should look like the one in *Figure 6.1*.

4. In this step, we will split the data into `training` and `testing`. We limit the testing size to 10% of the whole dataset. The length of the training set is 2002 and the length of the test set is 223:

```
bbc_train, bbc_test = train_test_split(bbc_df, test_size=0.1)
print(len(bbc_train))
print(len(bbc_test))
```

The result will be as follows:

```
2002
223
```

5. Here, we will assign the list of texts to the `documents` variable. We will then read in the `all-MiniLM-L6-v2` model to use for the sentence embeddings and encode the text data. Next, we will initialize a KMeans model with five clusters, setting the `n_init` parameter to `auto`, which determines the number of times the algorithm is run. We will also set the `init` parameter to `k-means++`. This parameter ensures faster convergence of the algorithm. We will then train the initialized model:

```
documents = bbc_train['text'].values
model = SentenceTransformer('all-MiniLM-L6-v2')
encoded_data = model.encode(documents)
km = KMeans(n_clusters=5, n_init='auto', init='k-means++')
km.fit(encoded_data)
```

6. Print out the most common words by topic:

```
print_most_common_words_by_cluster(documents, km, 5)
```

The results will vary; our results look like this:

```
0
['said', 'people', 'new', 'also', 'mr', 'technology', 'would',
'one', 'mobile', ...]
1
['said', 'game', 'england', 'first', 'win', 'world', 'last',
'would', 'one', 'two', 'time',...]
2
['said', 'film', 'best', 'music', 'also', 'year', 'us', 'one',
'new', 'awards', 'show',...]
3
['said', 'mr', 'would', 'labour', 'government', 'people',
'blair', 'party', 'election', 'also', 'minister', ...]
4
['said', 'us', 'year', 'mr', 'would', 'also', 'market',
'company', 'new', 'growth', 'firm', 'economy', ...]
```

We can see that the mapping of topics is as follows: 0 is tech, 1 is sport, 2 is entertainment, 3 is politics, and 4 is business.

7.   We can now use the test data to see how well the model performs on unseen data. First, we must create a prediction column in the test dataframe and populate it with the cluster number for each of the test inputs:

```
bbc_test["prediction"] = bbc_test["text"].apply(
    lambda x: km.predict(model.encode([x]))[0])
print(bbc_test)
```

The results will vary; this is our output:

```
          category                                         text \
578          sport   isinbayeva heads for birmingham olympic pole v...
2012 entertainment   boogeyman takes box office lead the low-budget...
208  entertainment   public show for reynolds portrait sir joshua r...
1078       politics   stalemate in pension strike talks talks aimed ...
1229         sport   newcastle 2-1 bolton kieron dyer smashed home ...
...            ...                                                ...
1977       politics   row over  police  power for csos the police fe...
1218       politics   whitehall cuts  ahead of target  thousands of ...
880  entertainment   roundabout continues nostalgia trip the new bi...
1552          tech   attack prompts bush site block the official re...
278       business   soaring oil  hits world economy  the soaring c...

     prediction
578           1
2012          2
208           2
1078          3
1229          1
...         ...
1977          3
1218          3
880           2
1552          0
278           4

[223 rows x 3 columns]
```

Figure 6.4 – The result of running K-Means on the test dataframe

8.   Now, we will create a mapping between the cluster number and the topic name, which we discovered manually by looking at the most frequent words for each cluster. We will then create a column with the predicted topic name for every text in the test set by using the mapping and the `prediction` column we created in the previous step. Now, we can compare the predictions of the model with the true value of the data. We will use the `classification_report` function from `sklearn` to get the corresponding statistics. Finally, we will print out the classification report for the predictions:

```
topic_mapping = {0:"tech", 1:"sport",
    2:"entertainment", 3:"politics", 4:"business"}
bbc_test["pred_category"] = bbc_test["prediction"].apply(
    lambda x: topic_mapping[x])
```

```
print(classification_report(bbc_test["category"],
    bbc_test["pred_category"]))
```

The result will be as follows:

	precision	recall	f1-score	support
business	0.98	0.96	0.97	55
entertainment	0.95	1.00	0.97	38
politics	0.97	0.93	0.95	42
sport	0.98	0.96	0.97	47
tech	0.93	0.98	0.95	41
accuracy			0.96	223
macro avg	0.96	0.97	0.96	223
weighted avg	0.96	0.96	0.96	223

The scores are very high – almost perfect. Most of this speaks to the quality of the sentence embedding model that we used.

9. Define a new example:

```
new_example = """Manchester United players slumped to the turf
at full-time in Germany on Tuesday in acknowledgement of what
their
latest pedestrian first-half display had cost them. The 3-2 loss
at
RB Leipzig means United will not be one of the 16 teams in the
draw
for the knockout stages of the Champions League. And this is not
the
only price for failure. The damage will be felt in the accounts,
in
the dealings they have with current and potentially future
players
and in the faith the fans have placed in manager Ole Gunnar
Solskjaer.
With Paul Pogba's agent angling for a move for his client and
ex-United
defender Phil Neville speaking of a "witchhunt" against his
former team-mate
Solskjaer, BBC Sport looks at the ramifications and reaction to
a big loss for United."""
```

10. Print the prediction for the new example:

```
predictions = km.predict(model.encode([new_example]))
print(predictions[0])
```

The output will be as follows:

```
1
```

Cluster number 1 corresponds to sport, which is the correct classification.

Topic modeling using BERTopic

In this recipe, we will explore the BERTopic package that provides many different and versatile tools for topic modeling and visualization. It is especially useful if you would like to do different visualizations of the topic clusters created. This topic modeling algorithm uses BERT embeddings to encode the data, hence the "BERT" in the name. You can learn more about the algorithm and its constituent parts at https://maartengr.github.io/BERTopic/algorithm/algorithm.html.

The BERTopic package, by default, uses the HDBSCAN algorithm to create clusters from the data in an unsupervised fashion. You can learn more about how the HDBSCAN algorithm works at https://hdbscan.readthedocs.io/en/latest/how_hdbscan_works.html. However, it is also possible to customize the inner workings of a BERTopic object to use other algorithms. It is also possible to substitute other custom components into its pipeline. In this recipe, we will use the default settings, and you can experiment with other components.

The resulting topics are of very high quality. There might be several reasons for this. One of them is the result of using BERT embeddings, which we saw in *Chapter 4*, to positively impact the classification results

Getting ready

We will use the BERTopic package to create topic models for the BBC dataset. The package is included in the poetry environment and is also part of the requirements.txt file.

The notebook is located at https://github.com/PacktPublishing/Python-Natural-Language-Processing-Cookbook-Second-Edition/blob/main/Chapter06/6.3-kmeans_with_bert.ipynb.

How to do it...

In this recipe, we will load the BBC dataset and preprocess it again. The preprocessing step will involve tokenizing the data and removing stopwords. Then we will create the topic model using BERTopic and inspect the results. We will also test the topic model on unseen data and use `classification_report` to see the accuracy statistics:

1. Do the necessary imports:

    ```
    import pandas as pd
    import numpy as np
    from bertopic import BERTopic
    from sklearn.model_selection import train_test_split
    from sklearn.metrics import classification_report
    ```

2. Run the language utilities file:

    ```
    %run -i "../util/lang_utils.ipynb"
    ```

3. Define and amend the stopwords, then read in the BBC data:

    ```
    stop_words = stopwords.words('english')
    stop_words.append("said")
    stop_words.append("mr")
    bbc_df = pd.read_csv("../data/bbc-text.csv")
    ```

    In this step, we will preprocess the data. We will first tokenize it using the `word_tokenize` method from NLTK as shown in the *Dividing sentences into words – tokenization* recipe in *Chapter 1*. Then remove stopwords and finally put the text back together into a string. We must do the last step because the BERTopic uses a sentence embedding model, and that model requires a string, not a list of words:

    ```
    bbc_df["text"] = bbc_df["text"].apply(
        lambda x: word_tokenize(x))
    bbc_df["text"] = bbc_df["text"].apply(
        lambda x: [w for w in x if w not in stop_words])
    bbc_df["text"] = bbc_df["text"].apply(lambda x: " ".join(x))
    ```

4. Here, we will split the dataset into training and test sets, specifying the size of the test set to be 10%. As a result, we will get 2002 datapoints for training and 223 for testing:

    ```
    bbc_train, bbc_test = train_test_split(bbc_df, test_size=0.1)
    print(len(bbc_train))
    print(len(bbc_test))
    ```

    The result will be as follows:

    ```
    2002
    223
    ```

5. Extract the lists of texts from the dataframe:

    ```
    docs = bbc_train["text"].values
    ```

6. In this step, we will initialize the `BERTopic` object and then fit it on the documents extracted in *step 6*. We will specify the number of topics to be six, one more than the five that we are looking for. This is because a key difference between BERTopic and other topic modeling algorithms is that it has a special **discard** topic numbered -1. We could also specify a larger number of topics. In that case, they would be narrower than the general five categories of business, politics, entertainment, tech, and sport:

```
topic_model = BERTopic(nr_topics=6)
topics, probs = topic_model.fit_transform(docs)
```

7. Here, we will print out the information about the resulting topic model. Other than the `discard` topic, the topics align well with the gold labels assigned by human annotators. The function prints out the most representative words for each topic, as well as the most representative documents:

```
print(topic_model.get_topic_info())
```

The results will vary; here is an example result:

```
   Topic  Count                             Name  \
0     -1    222             -1_also_company_china_us
1      0    463             0_england_game_win_first
2      1    393     1_would_labour_government_blair
3      2    321             2_film_best_music_awards
4      3    309  3_people_mobile_technology_software
5      4    294            4_us_year_growth_economy

                          Representation  \
0  [also, company, china, us, would, year, new, p...
1  [england, game, win, first, club, world, playe...
2  [would, labour, government, blair, election, p...
3  [film, best, music, awards, show, year, band, ...
4  [people, mobile, technology, software, digital...
5  [us, year, growth, economy, economic, company,...

                        Representative_Docs
0  [us retail sales surge december us retail sale...
1  [ireland win eclipses refereeing errors intern...
2  [lib dems unveil election slogan liberal democ...
3  [scissor sisters triumph brits us band scissor...
4  [mobiles media players yet mobiles yet ready a...
5  [consumer spending lifts us growth us economic...
```

8. In this step, we will print the topics. We can see from the words that the zeroth topic is sport, the first topic is politics, the second topic is entertainment, the third topic is tech, and the fourth topic is business.

```
print(topic_model.get_topic(0))
```

[('england', 0.023923609275000306), ('game', 0.023874910540888444), ('win', 0.02089139078895572), ('first', 0.019051267767033135), ('club', 0.017470682428724963), ('world', 0.017352529044283988), ('players', 0.01689906 2892940703), ('cup', 0.016893627391621816), ('last', 0.01665891773908297), ('two', 0.016416826185772164)]

```
print(topic_model.get_topic(1))
```

[('would', 0.035433209961128954), ('labour', 0.03291644460816451), ('government', 0.02936700096809325), ('blai r', 0.02702878364855685), ('election', 0.026944432307687366), ('party', 0.02566497689534048), ('people', 0.023 100026902541076), ('brown', 0.02021005312555692), ('minister', 0.020034108662757368), ('also', 0.0160661198584 5893)]

```
print(topic_model.get_topic(2))
```

[('film', 0.04996542271015715), ('best', 0.03954338090393644), ('music', 0.02608204595923062), ('awards', 0.02 2660883747149464), ('show', 0.019911582970118826), ('year', 0.019569114192597062), ('band', 0.0194726969350434 2), ('also', 0.01936668159798611), ('award', 0.019271695654981706), ('one', 0.018568435901997235)]

```
print(topic_model.get_topic(3))
```

[('people', 0.029907130205424383), ('mobile', 0.024988214218395956), ('technology', 0.02246648997447284), ('so ftware', 0.01907466920671406), ('digital', 0.0184666722908326), ('music', 0.018089961948297602), ('users', 0.01779304168540416), ('one', 0.017635710770530333), ('also', 0.017408837514157897), ('new', 0.017388243884138 55)]

```
print(topic_model.get_topic(4))
```

[('us', 0.03584149650622307), ('year', 0.02528335130307409), ('growth', 0.024390583399937417), ('economy', 0.0 21929580756366116), ('economic', 0.019792138393691337), ('company', 0.019580005023360762), ('yukos', 0.0189065 13989042615), ('market', 0.01801654052491897), ('oil', 0.017884696942644544), ('firm', 0.01642415825482552)]

Figure 6.5 – The topics generated by BERTopic

9. In this step, we will generate the topic labels using the `generate_topic_labels` function. We will input the number of words to use for the topic label, the separator (in this case, this is an underscore), and whether to include the topic number. As a result, we will get a list of topic names. We can see from the resulting topics that we could include *would* as a stopword:

```
topic_model.generate_topic_labels(
    nr_words=5, topic_prefix=True, separator='_')
```

The result will be similar to the following:

```
['-1_also_company_china_us_would',
 '0_england_game_win_first_club',
 '1_would_labour_government_blair_election',
 '2_film_best_music_awards_show',
 '3_people_mobile_technology_software_digital',
 '4_us_year_growth_economy_economic']
```

10. Here, we will define the `get_prediction` function that gives us the topic number for a text input and a corresponding model. The function transforms the input text and outputs a tuple of two lists. One is a list of topic numbers and the other is the list of probabilities of assigning each topic. The lists are sorted in the order of the most probable topic, so we can take the first element of the first list as the predicted topic and return it:

```
def get_prediction(input_text, model):
    pred = model.transform(input_text)
    pred = pred[0][0]
    return pred
```

11. In this step, we will define a column for predictions in the test dataframe and then use the function we defined in the previous step to get predictions for each text in the dataframe. We will then create a mapping of topic numbers to gold topic labels that we can use to test the effectiveness of the topic model:

```
bbc_test["prediction"] = bbc_test["text"].apply(
    lambda x: get_prediction(x, topic_model))
topic_mapping = {0:"sport", 1:"politics",
    2:"entertainment", 3:"tech", 4:"business", -1:"discard"}
```

12. Here, we will create a new column in the test dataframe to record the predicted topic name using the mapping we created. We will then filter the test set to only use entries that have not been predicted to be the discard topic -1:

```
bbc_test["pred_category"] = bbc_test["prediction"].apply(
    lambda x: topic_mapping[x])
test_data = bbc_test.loc[bbc_test['prediction'] != -1]
print(classification_report(test_data["category"],
    test_data["pred_category"]))
```

The result will be similar to this:

	precision	recall	f1-score	support
business	0.95	0.86	0.90	21
entertainment	0.97	1.00	0.98	30
politics	0.94	1.00	0.97	46
sport	1.00	1.00	1.00	62
tech	0.96	0.88	0.92	25
accuracy			0.97	184
macro avg	0.96	0.95	0.95	184
weighted avg	0.97	0.97	0.97	184

The test scores are very high. This is reflective of the encoding model, the BERTopic model, which is also a sentence transformer model, as in the previous recipe.

13. In this step, we will define a new example to test with the model and print it. We will use the `iloc` function from the `pandas` package to access the first element of the `bbc_test` dataframe:

```
new_input = bbc_test["text"].iloc[0]
print(new_input)
```

The result will be as follows (truncated):

```
howard dismisses tory tax fears michael howard dismissed fears
conservatives plans £4bn tax cuts modest . defended package
saying plan tories first budget hoped able go . tories monday
highlighted £35bn wasteful spending would stop allow tax cuts
reduced borrowing spending key services . ...
```

The example is about politics, which should be topic 1.

14. Obtain a prediction from the model and print it:

```
print(topic_model.transform(new_input))
```

The result will be a correct prediction:

```
([1], array([1.]))
```

There's more...

Now, we can find topics that are similar to a particular word, phrase, or sentence. This way, we could easily find a topic to which a text corresponds within the dataset. We will use a word, a phrase, and a sentence to see how well the model can show the corresponding topics:

1. Find topics most similar to the `sports` word with the corresponding similarity scores. Combine the topic numbers and similarity scores in a list of tuples and print them. The tuples are a combination of the topic number and the similarity score between the text that is passed in and the particular topic:

```
topics, similarity = topic_model.find_topics("sports", top_n=5)
sim_topics = list(zip(topics, similarity))
print(sim_topics)
```

The most similar topic is topic 0, which is sport:

```
[(0, 0.29033981040460977), (3, 0.049293092462828376), (-1,
-0.0047265937178774895), (2, -0.02074380026102955), (4,
-0.03699168959416969)]
```

2. Repeat the preceding step for the `business and economics` example phrase:

```
topics, similarity = topic_model.find_topics(
    "business and economics",
    top_n=5)
sim_topics = list(zip(topics, similarity))
print(sim_topics)
```

Here, the most similar topic is topic 4, which is business:

```
[(4, 0.29003573983158404), (-1, 0.26259758927249205),
(3, 0.15627005753581313), (1, 0.05491237184012845), (0,
0.010567363445904386)]
```

3. Now repeat the same process for the following example sentence: "YouTube removed
    a snippet of code that publicly disclosed whether a channel
    receives ad payouts, obscuring which creators benefit most from
    the platform. ". We would expect this to be most similar to the tech topic:

```
input_text = """YouTube removed a snippet of code that publicly
disclosed whether a channel receives ad payouts,
obscuring which creators benefit most from the platform."""
topics, similarity = topic_model.find_topics(
    input_text, top_n=5)
sim_topics = list(zip(topics, similarity))
print(sim_topics)
```

In the output, we can see that the most similar topic is topic 3, which is tech:

```
[(3, 0.2540850599909866), (-1, 0.172097560474608),
(2, 0.1367798346494483), (4, 0.10243553209139492), (1,
0.06954579004136925)]
```

Using contextualized topic models

In this recipe, we will look at another topic model algorithm: contextualized topic models. To produce a more effective topic model, it combines embeddings with a bag-of-words document representation.

We will show you how to use the trained topic model with input in other languages. This feature is especially useful because we can create a topic model in one language, for example, one that has many resources available, and then apply it on another language that does not have as many resources. To achieve this, we will utilize a multilingual embedding model in order to encode the data.

Getting ready

We will need the contextualized-topic-models package for this recipe. It is part of the poetry environment and the requirements.txt file.

The notebook is located at https://github.com/PacktPublishing/Python-Natural-Language-Processing-Cookbook-Second-Edition/blob/main/Chapter06/6.5-contextualized-tm.ipynb.

How to do it...

In this recipe, we will load the data, then divide it into sentences, preprocess it, and use the **gsdmm** model to cluster the sentences into topics. If you would like more information about the algorithm, please see the package documentation at https://pypi.org/project/contextualized-topic-models/.

1. Do the necessary imports:

```
import pandas as pd
from nltk.corpus import stopwords
from contextualized_topic_models.utils.preprocessing import(
    WhiteSpacePreprocessingStopwords)
from contextualized_topic_models.models.ctm import ZeroShotTM
from contextualized_topic_models.utils.data_preparation import(
    TopicModelDataPreparation)
```

2. Suppress the warnings:

```
import warnings
warnings.filterwarnings('ignore')
warnings.filterwarnings("ignore", category = DeprecationWarning)
import os
os.environ["TOKENIZERS_PARALLELISM"] = "false"
```

3. Create the stopwords list and read in the data:

```
stop_words = stopwords.words('english')
stop_words.append("said")
bbc_df = pd.read_csv("../data/bbc-text.csv")
```

4. In this step, we will create the preprocessor object and use it to preprocess the documents. The contextualized-topic-models package provides different preprocessors that prepare the data to be used in the topic model algorithm. This preprocessor tokenizes the documents, removes the stopwords, and puts them back into a string. It returns the list of preprocessed documents, the list of original documents, the dataset vocabulary, and a list of document indices in the original dataframe:

```
documents = bbc_df["text"]
preprocessor = WhiteSpacePreprocessingStopwords(
    documents, stopwords_list=stop_words)
preprocessed_documents,unpreprocessed_documents,vocab,indices =\
    preprocessor.preprocess()
```

5. Here, we will create the `TopicModelDataPreparation` object. We will pass the embedding model name as the parameter. This is a multilingual model that can encode text in various languages with good results. We will then fit it on the documents. It uses an embedding model to turn the texts into embeddings and also creates a bag-of-words model. The output is a `CTMDataset` object that represents the training dataset in the format required by the topic model training algorithm:

```
tp = TopicModelDataPreparation(
    "distiluse-base-multilingual-cased")
training_dataset = tp.fit(
    text_for_contextual=unpreprocessed_documents,
    text_for_bow=preprocessed_documents)
```

6. In this step, we will create the topic model using the `ZeroShotTM` object. The term **zero shot** means that the model has no prior information about the documents. We will input the size of the vocabulary for the bag-of-words model, the size of the embeddings vector, the number of topics (the `n_components` parameter), and the number of epochs to train the model for. We will use five topics, since the BBC dataset has that many topics. When you apply this algorithm to your data, you will need to experiment with different numbers of topics. Finally, we will fit the initialized topic model on the training dataset:

```
ctm = ZeroShotTM(bow_size=len(tp.vocab),
    contextual_size=512, n_components=5,
    num_epochs=100)
ctm.fit(training_dataset)
```

7. Here, we will inspect the topics. We can see that they fit well with the golden labels. Topic 0 is tech, topic 1 is sport, topic 2 is business, topic 3 is entertainment, and topic 4 is politics:

```
ctm.get_topics()
```

The results will vary; this is the output we get:

```
defaultdict(list,
            {0: ['people',       3: ['film',
                 'music',             'best',
                 'technology',        'actress',
                 'users',             'award',
                 'mobile',            'director',
                 'digital',           'awards',
                 'tv',                'stars',
                 'games',             'band',
                 'net',               'oscars',
                 'broadband'],        'album'],
             1: ['match',        4: ['mr',
                 'injury',            'government',
                 'side',              'labour',
                 'back',              'election',
                 'england',           'would',
                 'win',               'party',
                 'victory',           'blair',
                 'goal',              'minister',
                 'club',              'brown',
                 'great'],            'people']})
             2: ['analysts',
                 'bought',
                 'oil',
                 'figures',
                 'warned',
                 'value',
                 'securities',
                 'october',
                 'payments',
                 'analyst'],
```

Figure 6.6 – The contextualized model output

8. Now, we will initialize a new news piece, this time in Spanish, to see how effective the topic model trained on English-language documents will be on a news article in a different language. This particular news piece should fall into the tech topic. We will preprocess it using the `TopicModelDataPreparation` object. To then use the model on the encoded text, we need to create a dataset object. That is why we have to include the Spanish news piece in a list and then pass it on for data preparation. Finally, we must pass the dataset (that consists of only one element) through the model:

```
spanish_news_piece = """IBM anuncia el comienzo de la "era de la
utilidad cuántica" y anticipa un superordenador en 2033.
La compañía asegura haber alcanzado un sistema de computación
que no se puede simular con procedimientos clásicos."""
testing_dataset = tp.transform([spanish_news_piece])
```

9. In this step, we will get the topic distribution for the testing dataset we created in the previous step. The result is a list of lists, where each individual list represents the probability that a particular text belongs to that topic. The probabilities have the same indices in individual lists as the topic numbers:

```
ctm.get_doc_topic_distribution(testing_dataset)
```

In this case, the highest probability is for topic 0, which is indeed tech:

```
array([[0.5902461,0.09361929,0.14041995,0.07586181,0.0998529 ]],
      dtype=float32)
```

See also

For more information about contextualized topic models, see `https://contextualized-topic-models.readthedocs.io/en/latest/index.html`.

Visualizing Text Data

This chapter is dedicated to creating visualizations for the different aspects of NLP work, much of which we have done in previous chapters. Visualizations are important when working with NLP tasks, as they help us to easier see the big picture of the work accomplished.

We will create different types of visualizations, including visualizations of grammar details, parts of speech, and topic models. After working through this chapter, you will be well equipped to create compelling images to show and explain the outputs of various NLP tasks.

These are the recipes you will find in this chapter:

- Visualizing the dependency parse
- Visualizing parts of speech
- Visualizing NER
- Creating a confusion matrix plot
- Constructing word clouds
- Visualizing topics from Gensim
- Visualizing topics from BERTopic

Technical requirements

We will use the following packages in this chapter: `spacy`, `matplotlib`, `wordcloud`, and `pyldavis`. They are part of the `poetry` environment and the `requirements.txt` file.

We will be using two datasets in this chapter. The first is the BBC news dataset, located at `https://github.com/PacktPublishing/Python-Natural-Language-Processing-Cookbook-Second-Edition/tree/main/data/bbc_train.json` and `https://github.com/PacktPublishing/Python-Natural-Language-Processing-Cookbook-Second-Edition/tree/main/data/bbc_test.json`.

> **Note**
>
> This dataset is used in this book with permission from the researchers. The original paper associated with this dataset is as follows:
>
> Derek Greene and Pádraig Cunningham. "Practical Solutions to the Problem of Diagonal Dominance in Kernel Document Clustering," in Proc. 23rd International Conference on Machine Learning (ICML'06), 2006.
>
> All rights, including copyright, in the text content of the original articles are owned by the BBC.

The second is the Sherlock Holmes text, located at `https://github.com/PacktPublishing/Python-Natural-Language-Processing-Cookbook-Second-Edition/tree/main/data/sherlock_holmes.txt`.

Visualizing the dependency parse

In this recipe, we will learn how to use the `displaCy` library and visualize the dependency parse. It shows us the grammatical relations between words in a piece of text, usually a sentence.

Details about how to create a dependency parse can be found in *Chapter 2*, in the *Getting the dependency parse* recipe. We will create two visualizations, one for a short text and another for a long multi-sentence text.

After working through this recipe, you will be able to create visualizations of grammatical structures with different options for formatting.

Getting ready

The `displaCy` library is part of the `spacy` package. You need at least version 2.0.12 of the `spacy` package for `displaCy` to work. The version in the `poetry` environment and `requirements.txt` is 3.6.1.

The notebook is located at `https://github.com/PacktPublishing/Python-Natural-Language-Processing-Cookbook-Second-Edition/blob/main/Chapter07/7.1_dependency_parse.ipynb`.

How to do it...

To visualize the dependency parse, we will use the functionality of the `displaCy` package to first show one sentence, and then two sentences together:

1. Import the necessary packages:

   ```
   import spacy
   from spacy import displacy
   ```

2. Run the language utilities file:

```
%run -i "../util/lang_utils.ipynb"
```

3. Define the input text and process it using the small model:

```
input_text = "I shot an elephant in my pajamas."
doc = small_model(input_text)
```

4. We will now define different visualization options. The `add_lemma` option adds the word's lemma. For example, the lemma of *shot* is *shoot* and that is listed under the word itself. The `compact` option pushes the words and arrows together more, so the visualization fits in a smaller space. The `color` option changes the color of the words and arrows; for the options values, you can input either color names or color values in hex code. The `collapse_punct` option, if true, adds the punctuation to the word before it. The `arrow_spacing` option sets the distance between the arrows in pixels. The `bg` option sets the color of the background, whose value should be either a color name or a color code in hex. The `font` option changes the font of the words. The `distance` option sets the distance between words in pixels.

When we run the `render` command, we provide these options as an argument. We set the `jupyter` parameter to `True` for the visualization to work correctly in the notebook. You can omit the argument for non-Jupyter visualizations. We set the `style` parameter to `'dep'`, as we would like to have a dependency parse output. The output is a visual representation of the dependency parse:

```
options = {"add_lemma": True,
          "compact": True,
          "color": "green",
          "collapse_punct": True,
          "arrow_spacing": 20,
          "bg": "#FFFFE6",
          "font": "Times",
          "distance": 120}
display.render(doc, style='dep', options=options, jupyter=True)
```

The output is shown in *Figure 7.1*.

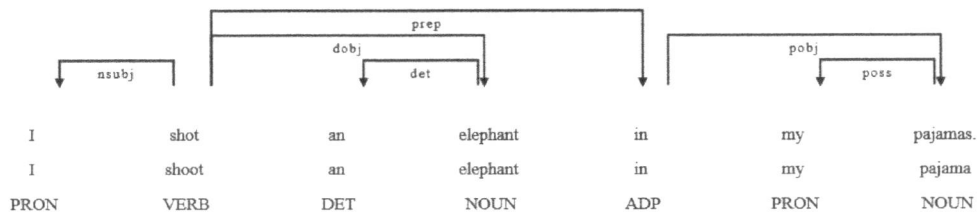

Figure 7.1 – Dependency parse visualization

1. In this step, we save the visualization to a file. We first import the `Path` object from the `pathlib` package. We then initialize a string with the path where we want to save the file and create a `Path` object. We use the same `render` command, this time saving the output in a variable and setting the `jupyter` parameter to `False`. We then use the `output_path` object and write the output to the corresponding file:

```
from pathlib import Path
path = "../data/dep_parse_viz.svg"
output_path = Path(path)
svg = displacy.render(doc, style="dep", jupyter=False)
output_path.open("w", encoding="utf-8").write(svg)
```

    This will create the dependency parse and save it at `../data/dep_parse_viz.svg`.

2. Now, let's define a longer text and process it using the small model. This way, we will be able to see how `displaCy` deals with longer texts:

```
input_text_list = "I shot an elephant in my pajamas. I hate it
    when elephants wear my pajamas."
doc = small_model(input_text_list)
```

3. Here, we visualize the new text. This time, we have to input a list of sentences from the processed `spacy` object to indicate that there is more than one sentence:

```
displacy.render(list(doc.sents), style='dep', options=options,
    jupyter=True)
```

The output should look like in *Figure 7.2*. We see that the output for the second sentence starts on a new line.

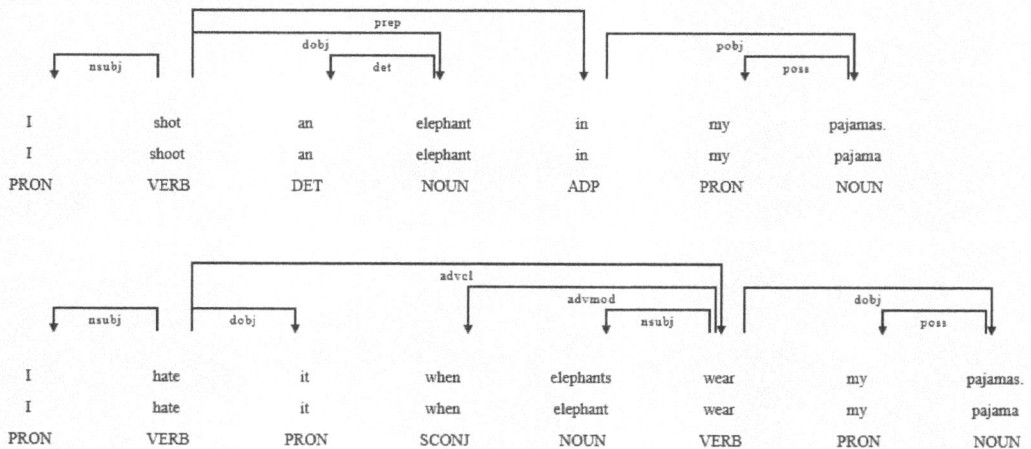

Figure 7.2 – Several sentences dependency parse visualization

Visualizing parts of speech

In this recipe, we visualize part of speech counts. Specifically, we count the number of infinitives and past or present verbs in the book *The Adventures of Sherlock Holmes*. This can give us an idea about whether the text mostly talks about past or present events. We could imagine that similar tools could be used to evaluate the quality of a text; for example, a book with very few adjectives but many nouns would not work very well as a fiction book.

After working through this recipe, you will be able to use the `matplotlib` package to create bar plots of different verb types, which are tagged using the `spacy` package.

Getting ready

We will use the `spacy` package for text analysis and the `matplotlib` package to create the graph. They are part of the `poetry` environment and the `requirements.txt` file.

The notebook is located at `https://github.com/PacktPublishing/Python-Natural-Language-Processing-Cookbook-Second-Edition/blob/main/Chapter07/7.2_parts_of_speech.ipynb`.

How to do it...

We will create a function that will count the number of verbs by tense and plot each on a bar graph:

1. Import the necessary packages:

    ```
    import spacy
    import matplotlib.pyplot as plt
    ```

2. Run the file and language utilities files. The language utilities notebook loads the `spacy` model, and the file utilities notebook loads the `read_text_file` function:

    ```
    %run -i "../util/lang_utils.ipynb"
    %run -i "../util/file_utils.ipynb"
    ```

3. Load the text of the Sherlock Holmes book:

    ```
    text_file = "../data/sherlock_holmes.txt"
    text = read_text_file(text_file)
    ```

4. Here, we define the verb tag lists, one for present tense and one for past tense. We do not define another list, but use it in the next step, and that is the infinitive verb, which only has one tag, VB. If you went through the *Part-of-speech tagging* recipe in *Chapter 1*, you will notice that the

tags are different from the `spacy` tags used there. These tags are more detailed and use the `tag_` attribute instead of the `pos_` attribute that is used in the simplified tagset:

```
past_tags = ["VBD", "VBN"]
present_tags = ["VBG", "VBP", "VBZ"]
```

5. In this step, we create the `visualize_verbs` function. The input to the function is the text and the `spacy` model. We check each token's `tag_` attribute and add the counts of present, past, and infinitive verbs to a dictionary. We then use the `pyplot` interface to plot those counts in a bar graph. We use the `bar` function to define the bar graph. The first argument lists the *x* coordinates of the bars. The next argument is a list of heights of the bars. We also set the `align` parameter to "center" and provide the colors for the bars using the `color` parameter. The `xticks` function sets the labels for the *x* axis. Finally, we use the `show` function to display the resulting plot:

```
def visualize_verbs(text, nlp):
    doc = nlp(text)
    verb_dict = {"Inf":0, "Past":0, "Present":0}
    for token in doc:
        if (token.tag_ == "VB"):
            verb_dict["Inf"] = verb_dict["Inf"] + 1
        if (token.tag_ in past_tags):
            verb_dict["Past"] = verb_dict["Past"] + 1
        if (token.tag_ in present_tags):
            verb_dict["Present"] = verb_dict["Present"] + 1
    plt.bar(range(len(verb_dict)),
        list(verb_dict.values()), align='center',
        color=["red", "green", "blue"])
    plt.xticks(range(len(verb_dict)),
        list(verb_dict.keys()))
    plt.show()
```

6. Run the `visualize_verbs` function on the text of the Sherlock Holmes book using the small `spacy` model:

```
visualize_verbs(text, small_model)
```

This will create the graph in *Figure 7.3*. We see that most of the verbs in the book are past tense, which makes sense for a novel. However, there is also a sizable number of present tense verbs, which could be part of direct speech.

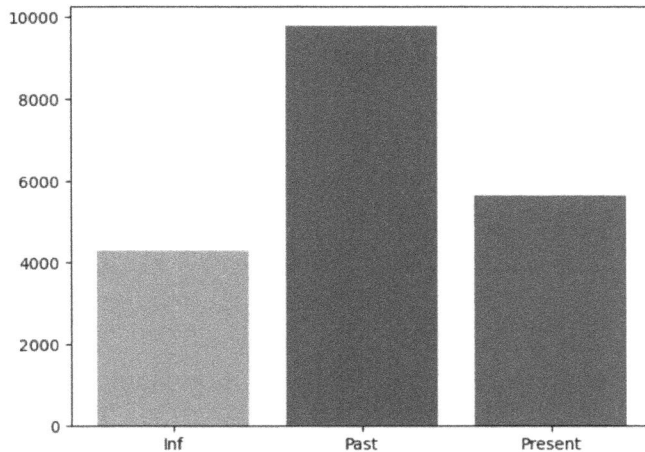

Figure 7.3 – Infinitive, past, and present verbs in The Adventures of Sherlock Holmes

Visualizing NER

Named entity recognition, or **NER**, is a very useful tool for quickly finding people, organizations, locations, and other entities in texts. In order to visualize them better, we can use the `displacy` package to create compelling and easy-to-read images.

After working through this recipe, you will be able to create visualizations of named entities in a text using different formatting options and save the results in a file.

Getting ready

The `displaCy` library is part of the `spacy` package. You need at least version 2.0.12 of the `spacy` package for `displaCy` to work. The version in the `poetry` environment and `requirements.txt` file is 3.6.1.

The notebook is located at `https://github.com/PacktPublishing/Python-Natural-Language-Processing-Cookbook-Second-Edition/blob/main/Chapter07/7.3_ner.ipynb`.

How to do it...

We will use `spacy` to parse the sentence and then the `displacy` engine to visualize the named entities:

1. Import both `spacy` and `displacy`:

    ```
    import spacy
    from spacy import displacy
    ```

2. Run the language utilities file:

```
%run -i "../util/lang_utils.ipynb"
```

3. Define the text to process:

```
text = """iPhone 12: Apple makes jump to 5G
Apple has confirmed its iPhone 12 handsets will be its first to
work on faster 5G networks.
The company has also extended the range to include a new "Mini"
model that has a smaller 5.4in screen.
The US firm bucked a wider industry downturn by increasing its
handset sales over the past year.
But some experts say the new features give Apple its best
opportunity for growth since 2014, when it revamped its line-up
with the iPhone 6.
"5G will bring a new level of performance for downloads and
uploads, higher quality video streaming, more responsive gaming,
real-time interactivity and so much more," said chief executive
Tim Cook.
There has also been a cosmetic refresh this time round, with the
sides of the devices getting sharper, flatter edges.
The higher-end iPhone 12 Pro models also get bigger screens than
before and a new sensor to help with low-light photography.
However, for the first time none of the devices will be bundled
with headphones or a charger."""
```

4. In this step, we process the text using the small model. This gives us a `Doc` object. We then modify the object to contain a title. This title will be part of the NER visualization:

```
doc = small_model(text)
doc.user_data["title"] = "iPhone 12: Apple makes jump to 5G"
```

5. Here, we set up color options for the visualization display. We set green for the ORG-labeled text and yellow for the PERSON-labeled text. We then set the `options` variable, which contains the colors. Finally, we use the `render` command to display the visualization. As arguments, we provide the `Doc` object and the options we previously defined. We also set the `style` argument to `"ent"`, as we would like to display just entities. We set the `jupyter` argument to `True` in order to display directly in the notebook:

```
colors = {"ORG": "green", "PERSON":"yellow"}
options = {"colors": colors}
displacy.render(doc, style='ent', options=options, jupyter=True)
```

The output should look like that in *Figure 7.4*.

iPhone 12: Apple makes jump to 5G

iPhone   12 **CARDINAL** :   Apple **ORG**   makes jump to   5   **CARDINAL**   G

Apple **ORG**   has confirmed its iPhone   12 **CARDINAL**   handsets will be its   first **ORDINAL**   to work on faster   5   **CARDINAL**   G networks.

The company has also extended the range to include a new "Mini" model that has a smaller   5.4 **CARDINAL**   in screen.

The   US **GPE**   firm bucked a wider industry downturn by increasing its handset sales over   the past year **DATE**   .

But some experts say the new features give   Apple **ORG**   its best opportunity for growth since   2014 **DATE**   , when it revamped its line-up with the iPhone 6.

"   5   **CARDINAL**   G will bring a new level of performance for downloads and uploads, higher quality video streaming, more responsive gaming,

real-time interactivity and so much more," said chief executive   Tim Cook **PERSON**   .

There has also been a cosmetic refresh this time round, with the sides of the devices getting sharper, flatter edges.

The higher-end iPhone   12 **CARDINAL**   Pro models also get bigger screens than before and a new sensor to help with low-light photography.

However, for the   first **ORDINAL**   time none of the devices will be bundled with headphones or a charger.

Figure 7.4 – Named entities visualization

6. Now we save the visualization to an HTML file. We first define the `path` variable. Then, we use the same `render` command, but we set the `jupyter` argument to `False` this time and assign the output of the command to the `html` variable. We then open the file, write the HTML, and close the file:

```
path = "../data/ner_vis.html"
html = displacy.render(doc, style="ent",
    options=options, jupyter=False)
html_file= open(path, "w", encoding="utf-8")
html_file.write(html)
html_file.close()
```

This will create an HTML file with the entities visualization.

Creating a confusion matrix plot

When working with machine learning models, for example, NLP classification models, creating a confusion matrix plot can be a very good tool to see the mistakes that the model makes to then further refine it. The model "confuses" one class for another, hence the name **confusion matrix**.

After working through this recipe, you will be able to create an SVM model, evaluate it, and then create a confusion matrix visualization that will tell you in detail which mistakes the model makes.

Getting ready

We will create an SVM classifier for the BBC news dataset using the sentence transformer model as the vectorizer. We will then use the `ConfusionMatrixDisplay` object to create a more informative confusion matrix. The classifier is the same as in the *Chapter 4* recipe *Using SVMs for supervised text classification.*

The dataset is located at `https://github.com/PacktPublishing/Python-Natural-Language-Processing-Cookbook-Second-Edition/tree/main/data/bbc_train.json` and `https://github.com/PacktPublishing/Python-Natural-Language-Processing-Cookbook-Second-Edition/tree/main/data/bbc_test.json`.

The notebook is located at `https://github.com/PacktPublishing/Python-Natural-Language-Processing-Cookbook-Second-Edition/blob/main/Chapter07/7.4_confusion_matrix.ipynb`.

How to do it...

1. Import the necessary packages and functions:

```
from sklearn.svm import SVC
from sentence_transformers import SentenceTransformer
from sklearn.metrics import confusion_matrix
import matplotlib.pyplot as plt
from sklearn.metrics import ConfusionMatrixDisplay
```

2. Run the simple classifier utilities file:

```
%run -i "../util/util_simple_classifier.ipynb"
```

3. Read in the training and test data and shuffle the training data. We shuffle the data so that there are no long sequences of one class, which might either bias the model during training or exclude large chunks of some classes:

```
train_df = pd.read_json("../data/bbc_train.json")
test_df = pd.read_json("../data/bbc_test.json")
train_df.sample(frac=1)
```

4. In this step, we load the transformer model and create the `get_sentence_vector` function. The function takes as arguments the text and the model, then creates and returns the vector. The `encode` method takes in a list of text, so in order to encode one piece of text, we need to put it into a list, and then get the first element of the `return` object, since the model also returns a list of encoding vectors:

```
model = SentenceTransformer('all-MiniLM-L6-v2')
def get_sentence_vector(text, model):
    sentence_embeddings = model.encode([text])
    return sentence_embeddings[0]
```

5. Here, we create the `train_classifier` function. The function takes in vectorized input and the correct answers. It then creates and trains an SVC object and returns it. It could take a few minutes to finish training:

```
def train_classifier(X_train, y_train):
    clf = SVC(C=0.1, kernel='rbf')
    clf = clf.fit(X_train, y_train)
    return clf
```

6. In this step, we train and test the classifier. First, we create a list with the target labels. We then create a `vectorize` function that uses the `get_sentence_vector` function but specifies the model to use. We then use the `create_train_test_data` function from the simple classifier utilities file to get the vectorized input and labels for both the training and test sets. This function takes in the training and test dataframes, the vectorizing method, and the name of the column where the text is located. The results are the vectorized training and test data and the true labels for both. Then, we use the `train_classifier` function to create a trained SVM classifier. We print the classification report for the training data and use the `test_classifier` function to print the classification report for the test data:

```
target_names=["tech", "business", "sport",
    "entertainment", "politics"]
vectorize = lambda x: get_sentence_vector(x, model)
(X_train, X_test, y_train, y_test) = create_train_test_data(
    train_df, test_df, vectorize,
    column_name="text_clean")
clf = train_classifier(X_train, y_train)
print(classification_report(train_df["label"],
        y_train, target_names=target_names))
test_classifier(test_df, clf, target_names=target_names)
```

The output should be as follows:

	precision	recall	f1-score	support
tech	1.00	1.00	1.00	321
business	1.00	1.00	1.00	408
sport	1.00	1.00	1.00	409
entertainment	1.00	1.00	1.00	309
politics	1.00	1.00	1.00	333
accuracy			1.00	1780
macro avg	1.00	1.00	1.00	1780
weighted avg	1.00	1.00	1.00	1780

	precision	recall	f1-score	support

tech	0.97	0.95	0.96	80
business	0.98	0.97	0.98	102
sport	0.98	1.00	0.99	102
entertainment	0.96	0.99	0.97	77
politics	0.98	0.96	0.97	84
accuracy			0.98	445
macro avg	0.97	0.97	0.97	445
weighted avg	0.98	0.98	0.98	445

7. Now, we create a mapping from number labels to text labels and then create a new column in the test dataframe that shows the text label prediction:

```
num_to_text_mapping = {0:"tech", 1:"business",
    2:"sport", 3:"entertainment", 4:"politics"}
test_df["pred_label"] = test_df["prediction"].apply(
    lambda x: num_to_text_mapping[x])
```

8. In this step, we create a confusion matrix using the sklearn confusion_matrix function. The function takes as input the true labels, the predictions, and the names of the categories. We then create a ConfusionMatrixDisplay object that takes in that confusion matrix and the names to display. We then create the confusion matrix plot using the object and display it using the matplotlib library:

```
cm = confusion_matrix(
    test_df["label_text"],
    test_df["pred_label"], labels=target_names)
disp = ConfusionMatrixDisplay(
    confusion_matrix=cm,
    display_labels=target_names)
disp.plot()
plt.show()
```

The result is shown in *Figure 7.5*. The resulting plot clearly shows which classes have overlaps and their number. For example, it is easy to see that there are two examples that are predicted to be about business but are actually about politics.

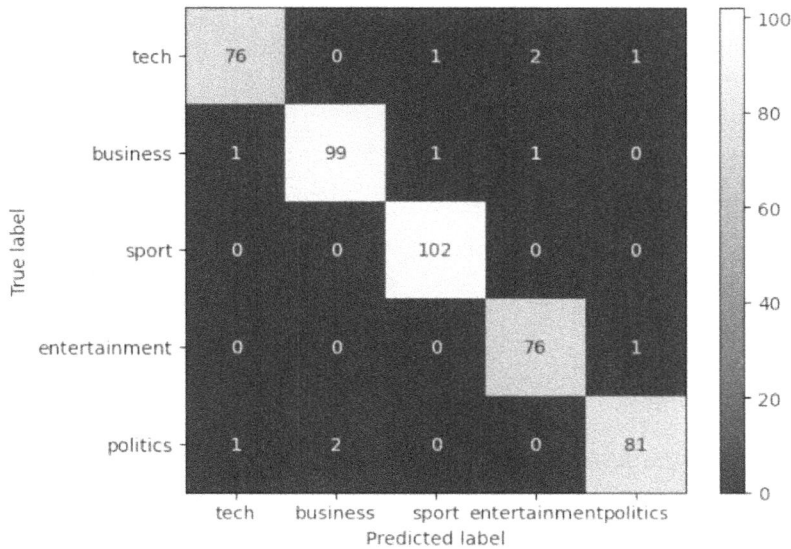

Figure 7.5 – Confusion matrix visualization

Constructing word clouds

Word clouds are a nice visualization tool to quickly see topics that are prevalent in a text. They can be used at the preliminary data analysis stage and for illustration purposes. A distinguishing feature of word clouds is that larger-font words signify a more frequent topic, while smaller-font words signify less frequent topics.

After working through this recipe, you will be able to create word clouds from a text and also apply a picture mask on top of the word cloud, which makes for a cool image.

We will use the text of the book *The Adventures of Sherlock Holmes* and the picture mask we will use is a silhouette of Sherlock Holmes' head.

Getting ready

We will use the `wordcloud` package for this recipe. In order to display the image, we need the `matplotlib` package as well. They are both part of the `poetry` environment and the `requirements.txt` file.

The notebook is located at `https://github.com/PacktPublishing/Python-Natural-Language-Processing-Cookbook-Second-Edition/blob/main/Chapter07/7.5_word_clouds.ipynb`.

How to do it...

1. Import the necessary packages and functions:

```
import matplotlib.pyplot as plt
from wordcloud import WordCloud, STOPWORDS
```

2. Run the file utilities notebook. We will use the `read_text_file` function from this notebook:

```
%run -i "../util/file_utils.ipynb"
```

3. Read in the book text:

```
text_file = "../data/sherlock_holmes.txt"
text = read_text_file(text_file)
```

4. In this step, we define the `create_wordcloud` function. The function takes as arguments the text to be processed, stopwords, the filename of where to save the result, and whether to apply a mask over the image (`None` by default). It creates the `WordCloud` object, saves it to the file, and then outputs the resulting plot. The options that we provide to the `WordCloud` object are the minimum font size, the maximum font size, the width and height, the maximum number of words, and the background color:

```
def create_wordcloud(text, stopwords, filename,
    apply_mask=None):
    if (apply_mask is not None):
        wordcloud = WordCloud(
            background_color="white", max_words=2000,
            mask=apply_mask, stopwords=stopwords,
            min_font_size=10, max_font_size=100)
        wordcloud.generate(text)
        wordcloud.to_file(filename)
        plt.figure()
        plt.imshow(wordcloud, interpolation='bilinear')
        plt.axis("off")
        plt.show()
    else:
        wordcloud = WordCloud(min_font_size=10,
            max_font_size=100, stopwords=stopwords,
            width=1000, height=1000, max_words=1000,
            background_color="white").generate(text)
        wordcloud.to_file(filename)
        plt.figure()
        plt.imshow(wordcloud, interpolation="bilinear")
        plt.axis("off")
        plt.show()
```

5. Run the `create_wordcloud` function on the text of the Sherlock Holmes book:

```
create_wordcloud(text, set(STOPWORDS),
    "../data/sherlock_wc.png")
```

This will save the result in the file located at `data/sherlock_wc.png` and create the visualization displayed in *Figure 7.6* (your results might look slightly different).

Figure 7.6 – Sherlock Holmes word cloud visualization

There's more...

We can also apply a mask to the word cloud. Here, we will apply a Sherlock Holmes silhouette to the word cloud:

1. Do the additional imports:

```
import numpy as np
from PIL import Image
```

2. Read in the mask image and save it as a numpy array:

```
sherlock_data = Image.open("../data/sherlock.png")
sherlock_mask = np.array(sherlock_data)
```

3. Run the function on the text of the Sherlock Holmes book:

```
create_wordcloud(text, set(STOPWORDS),
    "../data/sherlock_mask.png",
    apply_mask=sherlock_mask)
```

This will save the result in the file located at `data/sherlock_mask.png` and create the visualization shown in *Figure 7.7* (your result might be slightly different):

Figure 7.7 – Word cloud with mask

See also

Please see the `wordcloud` docs, `https://amueller.github.io/word_cloud/`, for more options.

Visualizing topics from Gensim

In this recipe, we will visualize the **Latent Dirichlet Allocation (LDA)** topic model that we created in *Chapter 6*. The visualization will allow us to quickly see words that are most relevant to a topic and the distances between topics.

After working through this recipe, you will be able to load an existing LDA model and create a visualization for its topics, both in Jupyter and saved as an HTML file.

Getting ready

We will use the pyLDAvis package to create the visualization. It is available in the poetry environment and the requirements.txt file.

We will load the model we created in *Chapter 6* and then use the pyLDAvis package to create the topic model visualization.

The notebook is located at https://github.com/PacktPublishing/Python-Natural-Language-Processing-Cookbook-Second-Edition/blob/main/Chapter07/7.6_topics_gensim.ipynb.

How to do it...

1. Import the necessary packages and functions:

    ```
    import gensim
    import pyLDAvis.gensim
    ```

2. Define the paths to the model files. The model was trained in *Chapter 6*:

    ```
    model_path = "../models/bbc_gensim/lda.model"
    dict_path = "../models/bbc_gensim/id2word.dict"
    corpus_path = "../models/bbc_gensim/corpus.mm"
    ```

3. In this step, we load the objects that these paths point to. If you get a FileNotFoundError error at this step, it means that you have not created the dictionary, corpus, and model files. In that case, go back to *Chapter 6*, the *LDA topic modeling with Gensim* recipe, and create the model and the accompanying files:

    ```
    dictionary = gensim.corpora.Dictionary.load(dict_path)
    corpus = gensim.corpora.MmCorpus(corpus_path)
    lda = gensim.models.ldamodel.LdaModel.load(model_path)
    ```

4. Here, we create the PreparedData object using the preceding files and save the visualization as HTML. The object is required for the visualization methods:

    ```
    lda_prepared = pyLDAvis.gensim.prepare(lda, corpus, dictionary)
    pyLDAvis.save_html(lda_prepared, '../data/lda-gensim.html')
    ```

5. Here, we enable the Jupyter display option and display the visualization in the notebook. You will see the topics and the words that are important for each topic. To select a particular

topic, hover over it with the mouse. You will see the most important words for each topic change while hovering over them:

```
pyLDAvis.enable_notebook()
pyLDAvis.display(lda_prepared)
```

This will create the visualization in *Figure 7.8* (your results might vary):

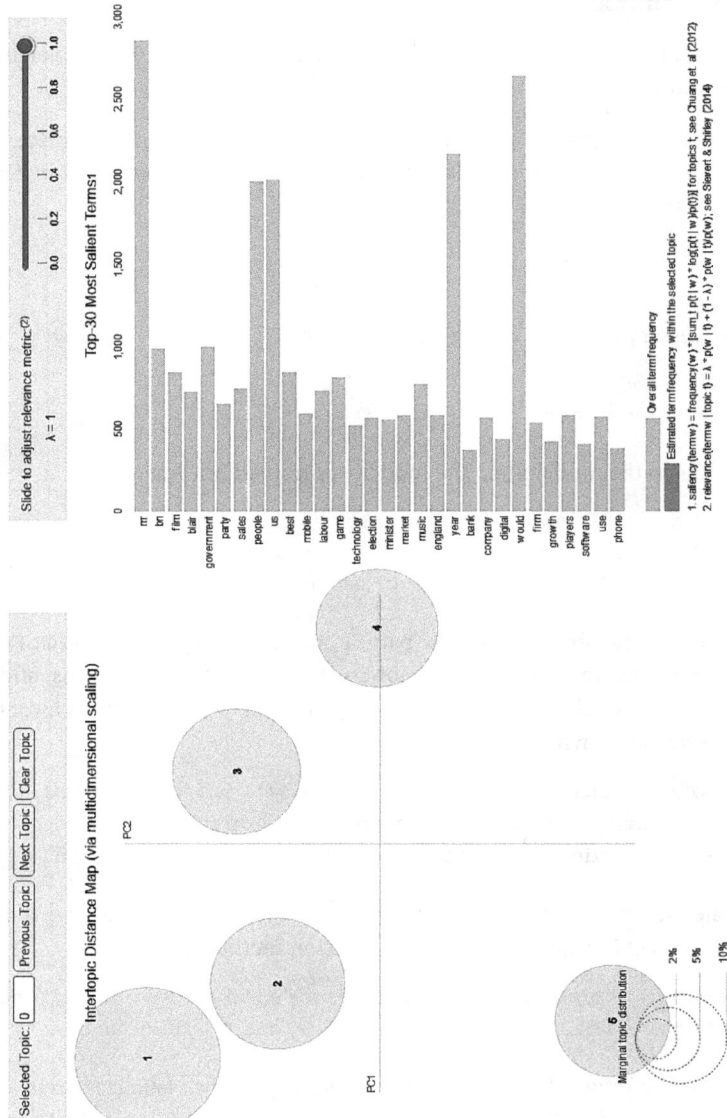

Figure 7.8 – LDA model visualization

See also

Using `pyLDAvis`, it is also possible to visualize models created using `sklearn`. See the package documentation for more information: `https://github.com/bmabey/pyLDAvis`.

Visualizing topics from BERTopic

In this recipe, we will create and visualize a BERTopic model on the BBC data. There are several visualizations available with the BERTopic package, and we will use several of them.

In this recipe, we will create a topic model in a similar fashion as in *Chapter 6*, in the *Topic modeling using BERTopic* recipe. However, unlike in *Chapter 6*, we will not limit the number of topics created, and resulting in more than the 5 original topics in the data. It will allow for more interesting visualizations.

Getting ready

We will use the `BERTopic` package to create the visualization. It is available in the `poetry` environment.

How to do it...

1. Import the necessary packages and functions:

```
import pandas as pd
import numpy as np
from bertopic import BERTopic
from bertopic.representation import KeyBERTInspired
```

2. Run the language utilities file:

```
%run -i "../util/lang_utils.ipynb"
```

3. Read in the data:

```
bbc_df = pd.read_csv("../data/bbc-text.csv")
```

4. Here, we create a list of training documents from the dataframe object. We then initialize a representation model object. Here, we use the `KeyBERTInspired` object, which uses BERT to extract the keywords.

   This object creates the names (representations) for the topics; it does a better job than the default version, which contains lots of stopwords. We then create the main topic model object and fit it to the document set. In this recipe, in contrast to the *Topic modeling using BERTopic* recipe in *Chapter 6*, we do not place limits on the number of created topics. This will create a lot more topics:

```
docs = bbc_df["text"].values
representation_model = KeyBERTInspired()
```

```
topic_model = BERTopic(
    representation_model=representation_model)
topics, probs = topic_model.fit_transform(docs)
```

5. In this step, we display the general topic visualization. It shows all 42 topics created. If you hover on each circle, you will see the topic representation or name. The representations consist of the top five words in the topic:

```
topic_model.visualize_topics()
```

This will create the visualization in *Figure 7.9* (your results might vary).

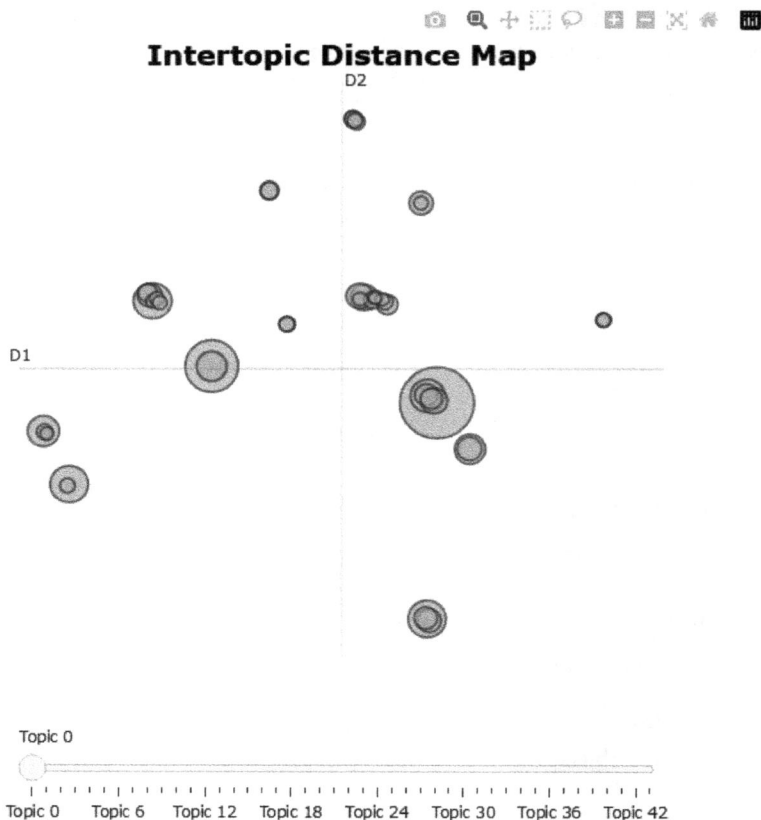

Figure 7.9 – BERTopic model visualization

6. Here, we create a visualization of the topic hierarchy. This hierarchy clusters the different topics together if they are related. We first create the hierarchy by using the `hierarchical_topics` function of the topic model object, and then pass it into the `visualize_hierarchy`

function. The nodes that combine the different topics have their own names that you can see if you hover over them:

```
hierarchical_topics = topic_model.hierarchical_topics(
    bbc_df["text"])
topic_model.visualize_hierarchy(
    hierarchical_topics=hierarchical_topics)
```

This will create the visualization in *Figure 7.10*.

Hierarchical Clustering

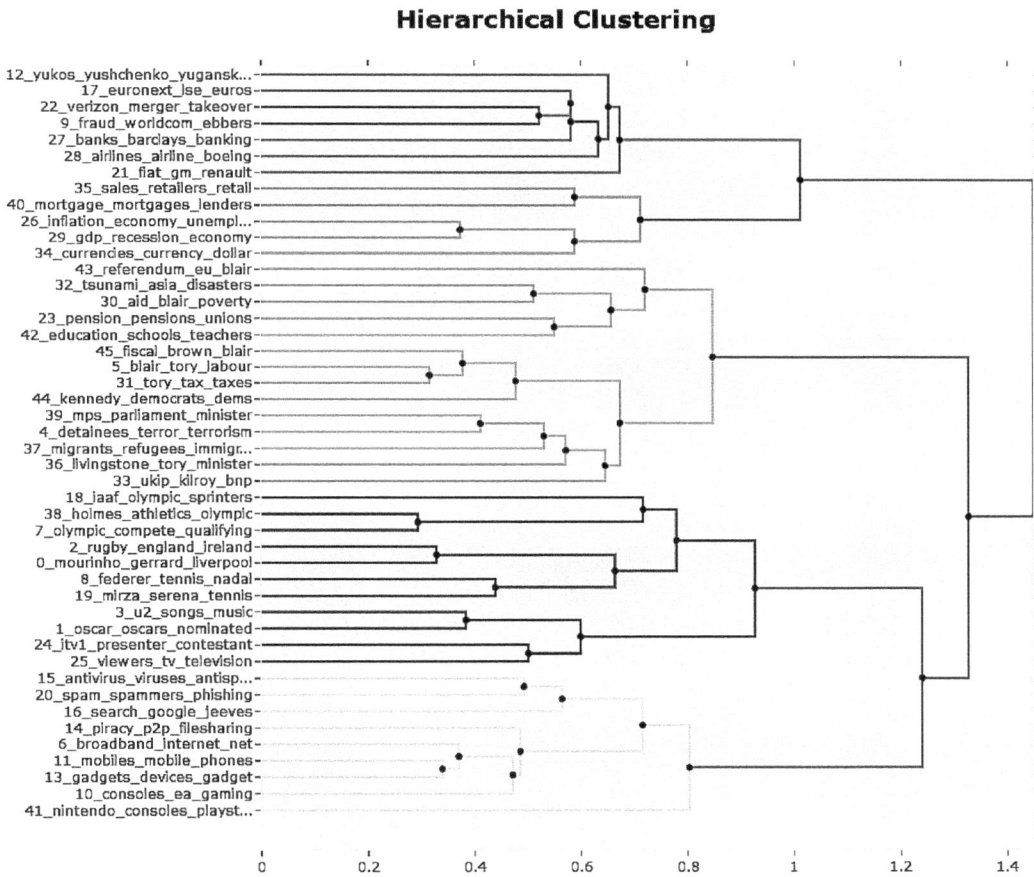

Figure 7.10 – BERTopic hierarchical visualization

If you hover over the nodes, you will see their names.

7.   In this step, we create a bar chart with the top words for the topics. We specify the number of topics to show by using the `top_n_topics` argument that the `visualize_barchart` function of the topic model object takes:

```
topic_model.visualize_barchart(top_n_topics=15)
```

This will create a visualization similar to this:

Topic Word Scores

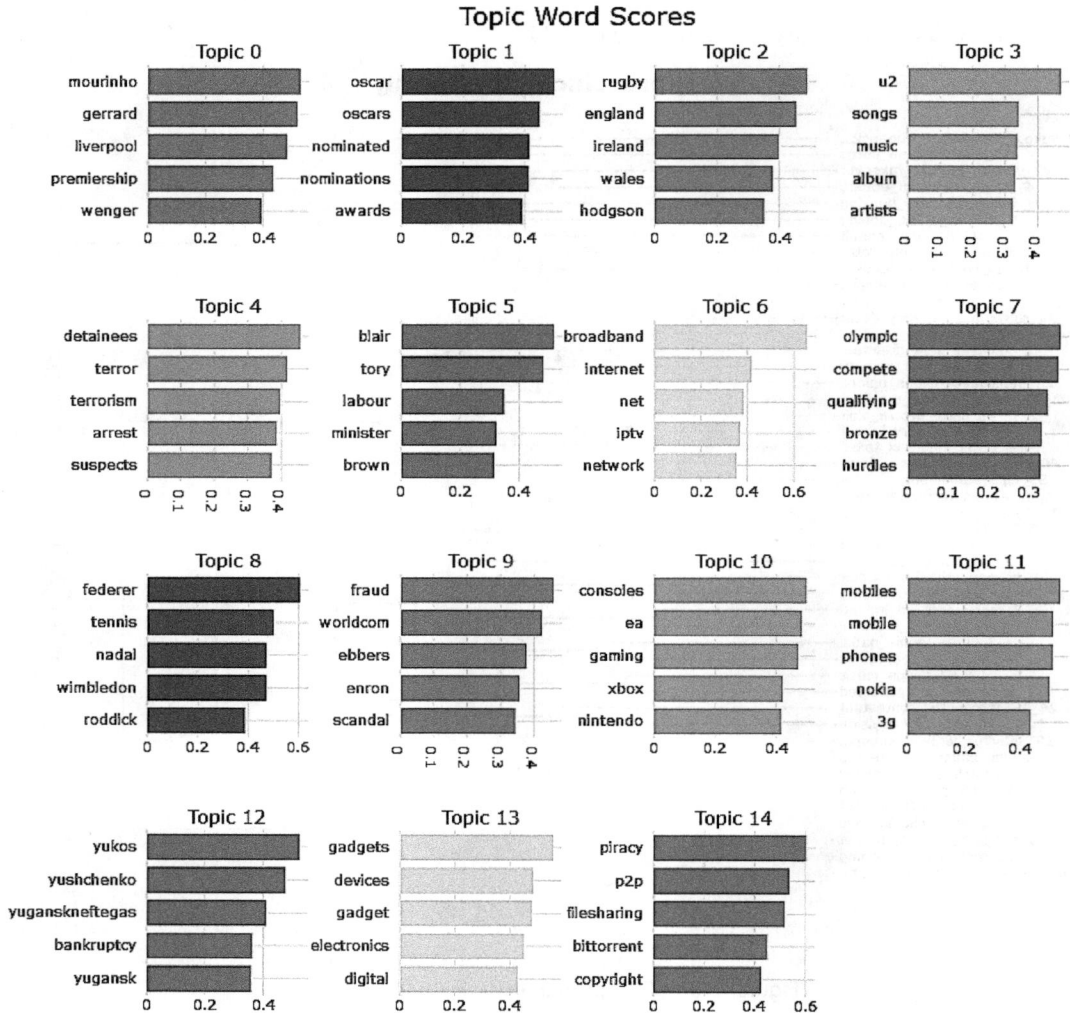

Figure 7.11 – BERTopic word scores

8. Here, we create a visualization of individual documents in the training set. We provide the list of documents created in *step 4* to the `visualize_documents` function. It clusters the documents to the topics. You can see the documents if you hover over the individual circles:

```
topic_model.visualize_documents(docs)
```

The result will be a visualization similar to this:

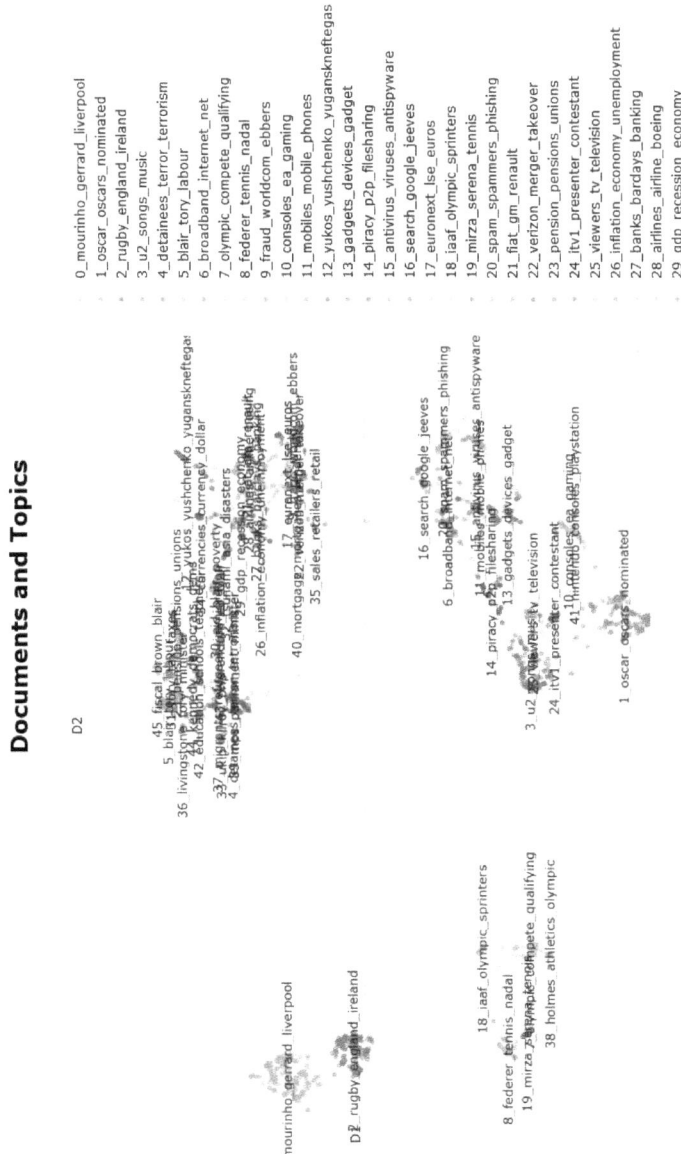

Figure 7.12 – BERTopic document visualization

If you hover over the nodes, you will see the text of the individual documents.

See also

- There are additional visualization tools available through BERTopic. See the package documentation for more information: `https://maartengr.github.io/BERTopic/index.html`.

- To learn more about `KeyBERTInspired`, see `https://maartengr.github.io/BERTopic/api/representation/keybert.html`.

8

Transformers and Their Applications

In this chapter, we will learn about transformers and how to apply them to perform various NLP tasks. Typical tasks in the NLP domain involve loading and processing data so that it can be used downstream seamlessly. Once the data is read, another task is that of transforming the data into a form that the various models can use. Once the data is transformed into the requisite format, we use it to perform the actual tasks, such as classification, text generation, and language translation.

Here is a list of the recipes in this chapter:

- Loading a dataset
- Tokenizing the text in your dataset
- Using the tokenized text to perform classification with Transformer models
- Using different Transformer models based on different requirements
- Generating text by taking a cue from an initial starting sentence
- Translating text between different languages using pre-trained Transformer models

Technical requirements

The code for this chapter is in the `Chapter08` folder in the GitHub repository of the book (`https://github.com/PacktPublishing/Python-Natural-Language-Processing-Cookbook-Second-Edition/tree/main/Chapter08`).

As in previous chapters, the packages required for the chapter are part of the `poetry` environment. Alternatively, you can install all the packages using the `requirements.txt` file.

Loading a dataset

In this recipe, we will learn how to load a public dataset and work with it. We will use the `RottenTomatoes` dataset for this recipe as an example. This dataset contains ratings and reviews for movies. Please refer to the link at `https://www.kaggle.com/datasets/stefanoleone992/rotten-tomatoes-movies-and-critic-reviews-dataset` for more information about the dataset

Getting ready

As part of this chapter, we will use the libraries from the `HuggingFace` site (`huggingface.co`). For this recipe, we will use the dataset package. You can use the `8.1_Transformers_dataset.ipynb` notebook from the code site if you need to work from an existing notebook.

How to do it...

In this recipe, you will load the `RottenTomatoes` dataset from the `HuggingFace` site using the dataset package. This package will download the dataset for you if it does not exist. For any subsequent runs, it will use the downloaded dataset from the cache if it was downloaded previously.

The recipe does the following things:

- Reads in the `RottenTomatoes` dataset
- Describes the features of the dataset
- Loads the data from the training split of the dataset
- Samples a few sentences from the dataset and prints them

The steps for the recipe are as follows:

1. Do the necessary imports to import the necessary types and functions from the datasets package:

    ```
    from datasets import load_dataset, get_dataset_split_names
    ```

2. Load `"rotten tomatoes"` via the `load_dataset` function and print the internal dataset splits. This dataset contains the train, validation, and test splits:

    ```
    dataset = load_dataset("rotten_tomatoes")
    print(get_dataset_split_names("rotten_tomatoes"))
    ```

    The output of the preceding command would be as follows:

    ```
    ['train', 'validation', 'test']
    ```

3. Load the dataset and print the attributes of the train split. The `training_data.description` describes the dataset details and the `training_data.features` describes the features of the dataset. In the output, we can see that the `training_data` split contains the features `text`, which is of the type string, and `label`, which is of type categorical, with the values `neg` and `pos`:

```
training_data = dataset['train']
print(training_data.description)
print(training_data.features)
```

The output of the command is as follows:

```
Movie Review Dataset.
This is a dataset of containing 5,331 positive and 5,331
negative processed  sentences from Rotten Tomatoes movie
reviews. This data was first used in Bo Pang and Lillian Lee,
``Seeing stars: Exploiting class relationships for sentiment
categorization with respect to rating scales.'', Proceedings of
the ACL, 2005.

{'text': Value(dtype='string', id=None),
    'label':ClassLabel(names=['neg', 'pos'], id=None)}
```

4. Now that we have loaded the dataset, we will print the first five sentences from it. This is just to confirm that we are indeed able to read from the dataset:

```
sentences = training_data['text'][:5]
[print(sentence) for sentence in sentences]
```

The output of the command is as follows:

```
the rock is destined to be the 21st century's new " conan "
and that he's going to make a splash even greater than arnold
schwarzenegger , jean-claud van damme or steven segal .

the gorgeously elaborate continuation of " the lord of the rings
" trilogy is so huge that a column of words cannot adequately
describe co-writer/director peter jackson's expanded vision of j
. r . r . tolkien's middle-earth .

effective but too-tepid biopic

if you sometimes like to go to the movies to have fun , wasabi
is a good place to start .

emerges as something rare , an issue movie that's so honest and
keenly observed that it doesn't feel like one .
```

Tokenizing the text in your dataset

The components contained within the transformer do not have any intrinsic knowledge of the words that it processes. Instead, the tokenizer only uses the token identifiers for the words that it processes. In this recipe, we will learn how to transform the text in your dataset into a representation that can be used by the models for downstream tasks.

Getting ready

As part of this recipe, we will use the `AutoTokenizer` module from the transformers package. You can use the `8.2_Basic_Tokenization.ipynb` notebook from the code site if you need to work from an existing notebook.

How to do it...

In this recipe, you will continue from the previous example of using the `RottenTomatoes` dataset and sampling a few sentences from it. We will then encode the sampled sentences into tokens and their respective representations.

The recipe does the following things:

- Loads a few sentences into memory
- Instantiates a tokenizer and tokenizes the sentences
- Converts the token IDs generated from the previous step back into the tokens

The steps for the recipe are as follows:

1. Do the necessary imports to import the necessary `AutoTokenizer` module from the `transformers` library:

    ```
    from transformers import AutoTokenizer
    ```

2. We initialize a sentence array consisting of three sentences that we will use for this example. These sentences are of different lengths and have a good combination of the same and different words. This will allow us to understand how the tokenized representation varies for each of them:

    ```
    sentences = [
        "The first sentence, which is the longest one in the list.",
        "The second sentence is not that long.",
        "A very short sentence."]
    ```

3. Instantiate a tokenizer of the `bert-base-cased` type. This tokenizer is case-sensitive. This means that the words star and STAR will have different tokenized representations:

    ```
    tokenizer = AutoTokenizer.from_pretrained("bert-base-cased")
    ```

4. In this step, we tokenize all the sentences in the `sentences` array. We call the tokenizer constructor and pass it the `sentences` array as an argument followed by printing the `tokenized_output` instance returned by the constructor function. This object is a dictionary of three items:

    - `input_ids`: These are the numerical token identifiers that are assigned to each token.

    - `token_type_ids`: These IDs define the type of tokens that are contained in the sentences.

    - `attention_mask`: These define the attention values for each token in the input. This mask determines what tokens are paid attention to when downstream tasks are performed. These values are floats and can vary from 0 (no attention) to 1 (full attention).

    ```
    tokenized_input = tokenizer(sentences)
    print(tokenized_input)
    ```

    ```
    {'input_ids': [[101, 1109, 1148, 5650, 117, 1134, 1110, 1103,
    6119, 1141, 1107, 1103, 2190, 119, 102],
    [101, 1109, 1248, 5650, 1110, 1136, 1115, 1263, 119, 102],[101,
    138, 1304, 1603, 5650, 119, 102]],
    'token_type_ids': [[0, 0, 0, 0, 0, 0, 0, 0, 0, 0, 0, 0, 0, 0,
    0], [0, 0, 0, 0, 0, 0, 0, 0, 0, 0], [0, 0, 0, 0, 0, 0, 0]],
    'attention_mask': [[1, 1, 1, 1, 1, 1, 1, 1, 1, 1, 1, 1, 1, 1,
    1], [1, 1, 1, 1, 1, 1, 1, 1, 1, 1], [1, 1, 1, 1, 1, 1, 1]]}
    ```

5. In this step, we take the input IDs of the first sentence and convert them back into tokens:

    ```
    tokens = tokenizer.convert_ids_to_tokens(
        tokenized_input["input_ids"][0])
    print(tokens)
    ```

    ```
    ['[CLS]', 'The', 'first', 'sentence', ',', 'which', 'is', 'the',
    'longest', 'one', 'in', 'the', 'list', '.', '[SEP]']
    ```

    The token IDs for the first sentence are in the following form:

    ```
    [101, 1109, 1148, 5650, 117, 1134, 1110, 1103, 6119, 1141, 1107,
    1103, 2190, 119, 102]
    ```

    Converting them into tokens returns the following output:

    ```
    ['[CLS]', 'The', 'first', 'sentence', ',', 'which', 'is', 'the',
    'longest', 'one', 'in', 'the', 'list', '.', '[SEP]']
    ```

In addition to the original tokens, the tokenizer adds the **classification** (**CLS**) and the **separator** (**SEP**) tokens to the beginning and the end of the sentence, respectively, denoted as [CLS] and [SEP]. These tokens were added for the training tasks that were performed to train BERT.

Now that we have learned about the internal representation of the text used by the transformer internally, let us learn how we can classify a piece of text into different categories.

Classifying text

In this recipe, we will use the RottenTomatoes dataset and classify the review texts for sentiment. We will classify the test split of the dataset and evaluate the results of the classifier against the true labels in the test split of the dataset.

Getting ready

As part of this recipe, we will use the pipeline module from the transformers package. You can use the 8.3_Classification_And_Evaluation.ipynb notebook from the code site if you need to work from an existing notebook.

How to do it...

In this recipe, you will continue from the previous example using the RottenTomatoes dataset and sample a few sentences from it. We will then classify a small subset of five sentences for sentiment classification and demonstrate the results on this smaller subset. We will then perform inference on the whole test split of the dataset and evaluate the results of the classification.

The recipe does the following things:

- Loads the RottenTomatoes dataset and prints the first five sentences from it
- Instantiates a pipeline with a pre-trained Roberta model that has been trained on the same dataset for sentiment analysis
- Performs inference (or sentiment prediction) on the whole test split of the dataset using the pipeline
- Evaluates the results of the inference

The steps for the recipe are as follows:

1. Do the necessary imports to import the required packages and modules:

    ```
    from datasets import load_dataset
    from evaluate import evaluator, combine
    from transformers import pipeline
    import torch
    ```

2. In this step, we probe for the presence of a **Compute Unified Device Architecture (CUDA)** compatible device (or **Graphics Processing Unit (GPU)**) present in the system. If such a device is present, our model will be loaded on it. This accelerates the training and inference performance of a model if it is supported. However, if such a device is not present, the **Central Processing**

Unit (**CPU**) will be used. We also load the `RottenTomatoes` dataset and select the first five sentences from it. This is to ensure we are indeed able to read the data present in the dataset:

```
device = torch.device(
    "cuda" if torch.cuda.is_available() else "cpu")
sentences = load_dataset(
    "rotten_tomatoes", split="test").select(range(5))
[print(sentence) for sentence in sentences['text']]
lovingly photographed in the manner of a golden book sprung
to life , stuart little 2 manages sweetness largely without
stickiness .

consistently clever and suspenseful .

it's like a " big chill " reunion of the baader-meinhof gang
, only these guys are more harmless pranksters than political
activists .

the story gives ample opportunity for large-scale action and
suspense , which director shekhar kapur supplies with tremendous
skill .

red dragon " never cuts corners .
```

3. Initialize the pipeline for sentiment analysis via a pipeline. A pipeline is an abstraction that allows us to easily use models or inference tasks without having to write the code to piece them together. We load the `roberta-base-rotten-tomatoes` model from `textattack`, which has been trained on this dataset. In the following segment, we use the pipeline for the sentiment analysis task and set a specific model to be used for this task:

```
roberta_pipe = pipeline("sentiment-analysis",
    model="textattack/roberta-base-rotten-tomatoes")
```

4. In this step, we generate the predictions for the small subset of sentences that we selected in step 2. Using the pipeline object to generate predictions is as easy as just passing it a series of sentences. If you are running this example on a machine without a compatible CUDA device, this step might take a little time:

```
predictions = roberta_pipe(sentences['text'])
```

5. In this step, we iterate through our sentences and check the predictions for our sentences. We print the actual and generated predictions along with the sentence text for the five sentences. The actual labels are read from the dataset, whereas the predictions are generated via the pipeline object:

```
for idx, _sentence in enumerate(sentences['text']):
    print(
```

```
                    f"actual: {sentences['label'][idx]}\n"
                    f"predicted: {'1' if predictions[idx]['label']
                        == 'LABEL_1' else '0'}\n"
                    f"sentence: {_sentence}\n\n"
            )
```

```
actual:1
predicted:1
sentence:lovingly photographed in the manner of a golden book
sprung to life , stuart little 2 manages sweetness largely
without stickiness .

actual:1
predicted:1
sentence:consistently clever and suspenseful .

actual:1
predicted:0
sentence:it's like a " big chill " reunion of the baader-
meinhof gang , only these guys are more harmless pranksters than
political activists .

actual:1
predicted:1
sentence:the story gives ample opportunity for large-scale
action and suspense , which director shekhar kapur supplies with
tremendous skill .

actual:1
predicted:1
sentence:red dragon " never cuts corners .
```

6. Now that we have validated the pipeline and its results, let's generate the inference for the whole test set and generate the evaluation measures of this particular model. Load the complete test split for the `RottenTomatoes` dataset:

    ```
    sentences = load_dataset("rotten_tomatoes", split="test")
    ```

7. In this step, we initialize an evaluator object that can be used to perform the inference along with evaluating the results of the classification. It can also be used to present an easy-to-read summary of the evaluation results:

    ```
    task_evaluator = evaluator("sentiment-analysis")
    ```

8. In this step, we call the `compute` method on the `evaluator` instance. This triggers the inference and the evaluation using the same pipeline instance that we initialized in step 4. It returns the evaluation metrics of `accuracy`, `precision`, `recall`, and `f1`, along with some performance metrics related to inference:

```
eval_results = task_evaluator.compute(
    model_or_pipeline=roberta_pipe,
    data=sentences,
    metric=combine(["accuracy", "precision", "recall", "f1"]),
    label_mapping={"LABEL_0": 0, "LABEL_1": 1}
)
```

9. In this step, we print the results of the evaluation. Of note are the `precision`, `recall`, and `f1` values. An `f1` of `0.88`, observed in this case, is an indicator of the very good efficacy of the classifier, though it could always be improved further:

```
print(eval_results)
{'accuracy': 0.88,
 'precision': 0.92,
 'recall': 0.84,
 'f1': 0.88,
 'total_time_in_seconds': 27.23,
 'samples_per_second': 39.146,
 'latency_in_seconds': 0.025}
```

In this recipe, we used a pre-trained classifier to classify the data on a dataset. The dataset and model were both for sentiment analysis. There are cases where we can use classifiers that are trained on a different class of data but can still be used as is. This saves us from having to train a classifier of our own and repurpose a model that already exists. We will learn about this use case in the next recipe.

Using a zero-shot classifier

In this recipe, we will classify a sentence using a zero-shot classifier. There are instances where we do not have the luxury of training a classifier from scratch or using a model that has been trained as per the labels of our data. **Zero-shot classification** can be used in such scenarios for any team to get up and running quickly. The zero in the terminology means that the classifier has not seen any data (zero samples precisely) from the target dataset that will be used for inference.

Getting ready

As part of this recipe, we will use the pipeline module from the transformers package. You can use the `8.4_Zero_shot_classification.ipynb` notebook from the code site if you need to work from an existing notebook.

How to do it...

In this recipe, we will use a couple of sentences and classify them. We will use our own set of labels for these sentences. We will use the `facebook/bart-large-mnli` model for this recipe. This model is suitable for the task of zero-shot classification.

The recipe does the following things:

- Initializes a pipeline based on a zero-shot classification model
- Uses the pipeline to classify a sentence into a custom set of user-defined labels
- Prints the results of the classification with the classes and their associated probabilities

The steps for the recipe are as follows:

1. Do the necessary imports and identify the compute device, as described in the previous classification recipe:

```
from transformers import pipeline
import torch
device = torch.device(
    "cuda" if torch.cuda.is_available() else "cpu")
```

2. In this step, we initialize a pipeline instance with the `facebook/bart-large-mnli` model. We have chosen this particular model for our example, but other models can also be used – available on the `HuggingFace` site:

```
pipeline_instance = pipeline(
    model="facebook/bart-large-mnli")
```

3. Use the pipeline instance to classify a sentence into a given set of candidate labels. The labels provided in the example are completely novel and have been defined by us. The model was not trained on examples with these labels. The classification output is stored in the `result` variable, which is a dictionary. This dictionary has the `'sequence'`, `'labels'`, and `'scores'` keys. The `'sequence'` element stores the original sentence passed to the classifier. The `'labels'` element stores the labels for the classes, but the ordering is different than what we passed in the arguments. The `'scores'` element stores the probabilities of the classes and corresponds to the same ordering in the `'labels'` element. The last argument in this call is `device`. If there is a CUDA-compatible device present in the system, it will be used:

```
result = pipeline_instance(
    "I am so hooked to video games as I cannot get any work
done!",
    candidate_labels=["technology", "gaming", "hobby", "art",
"computer"], device=device)
```

4. We print the sequence, followed by printing each label and its associated probability. Note that the order of the labels has changed from the initial input that we specified in the previous step. The function call reorders the labels based on the descending order of the label probability:

```
print(result['sequence'])
for i, label in enumerate(result['labels']):
    print(f"{label}:  {result['scores'][i]:.2f}")
I am so hooked to video games as I cannot get any work done!
gaming:  0.85
hobby:  0.08
technology:  0.07
computer:  0.00
art:  0.00
```

5. We run a zero-shot classification on a different sentence and print the results for it. This time, we emit a result that picks the class with the highest probability and prints the result:

```
result = pipeline_instance(
    "A early morning exercise regimen can drive many diseases
away!",
    candidate_labels=["health", "medical", "weather",
"geography", "politics"], )

print(result['sequence'])
for i, label in enumerate(result['labels']):
    print(f"{label}:  {result['scores'][i]:.2f}")
print(
    f"The most probable class for the sentence is **
    {result['labels'][0]} ** "
    f"with a probability of {result['scores'][0]:.2f}"
)
A early morning exercise regimen can drive many diseases away!
health:  0.91
medical:  0.07
weather:  0.01
geography:  0.01
politics:  0.00
The most probable class for the sentence is ** health ** with a
probability of 0.91
```

So far, we have used the transformer and some pre-trained models to generate token IDs and classifications. These recipes have used the encoder part of the transformer. The encoder generates a representation of the text, which is then used by a classifier head in front of it to generate classification labels. However, the transformer has another component, called the decoder. A decoder uses a given representation of text and generates subsequent text. In the next recipe, we will learn more about the decoder.

Generating text

In this recipe, we will use a **generative transformer model** to generate text from a given seed sentence. One such model to generate text is the GPT-2 model, which is an improved version of the original **General Purpose Transformer (GPT)** model.

Getting ready

As part of this recipe, we will use the pipeline module from the transformers package. You can use the 8.5_Transformer_text_generation.ipynb notebook from the code site if you need to work from an existing notebook.

How to do it...

In this recipe, we will start with an initial seed sentence and use the GPT-2 model to generate text based on the given seed sentence. We will also tinker with certain parameters to improve the quality of the generated text.

The recipe does the following things:

- It initializes a starting sentence from which a continuing sentence will be generated
- It initializes a GPT-2 model as part of a pipeline and uses it to generate five sentences as part of the parameters passed to the generator method
- It prints the results of the generation

The steps for the recipe are as follows:

1. Do the necessary imports and identify the compute device, as described in the previous classification recipe:

```
from transformers import pipeline
import torch
device = torch.device(
    "cuda" if torch.cuda.is_available() else "cpu")
```

2. Initialize a seed input sentence based on which the subsequent text will be generated. Our goal here is to use the GPT-2 decoder to hypothetically generate the text that follows it based on the generation parameters:

```
text = "The cat had no business entering the neighbors garage,
but"
```

3. In this step, we initialize a text-generation pipeline with the `'gpt-2'` model. This model is based on a **large language model** (**LLM**) that was trained using a large text corpus. The last argument in this call is `device`. If there is a CUDA-compatible device present in the system, it will be used:

```
generator = pipeline(
    'text-generation', model='gpt2', device=device)
```

4. Generate the continuing sequence for the seed sentence and store the results. The parameters of note used in the call other than the seed text are as follows:

   - `max_length`: The maximum length of the generated sentence, including the length of the seed sentence.

   - `num_return_sequences`: The number of generated sequences to return.

   - `num_beams`: This parameter controls the quality of the generated sequence. A higher number generally results in improved quality of the generated sequence but also slows down the generation. We encourage you to try out different values for this parameter based on the quality requirements of the generated sequence.

```
generated_sentences = generator(
    text,do_sample=True, max_length=30,
    num_return_sequences=5, num_beams=5,
    pad_token_id=50256)
```

5. Print the generated sentences:

```
[print(generated_sentence['generated_text'])
    for generated_sentence in generated_sentences]


The cat had no business entering the neighbors garage, but  he
was able to get inside.  The cat had been in the neighbor's
The cat had no business entering the neighbors garage, but  the
owner of the house called 911.  He said he found the cat in
The cat had no business entering the neighbors garage, but  he
was able to get his hands on one of the keys.  It was
The cat had no business entering the neighbors garage, but  he
didn't seem to mind at all.  He had no idea what he
The cat had no business entering the neighbors garage, but  the
cat had no business entering the neighbors garage, but the cat
had no business entering
```

There's more...

As we can see in the preceding example, the generated output was rudimentary, repetitive, grammatically incorrect, or perhaps incoherent. There are different techniques that we can use to improve the generated output.

We will use the `no_repeat_ngram_size` parameter this time to generate the text. We will set the value of this parameter to 2. This instructs the generator to not repeat bi-grams.

We will change the line in *step 4* to the following:

```
generated_sentences = generator(text, do_sample=True,
    max_length=30, num_return_sequences=5, num_beams=5,
    no_repeat_ngram_size=2,  pad_token_id=50256)
```

As we can see in the following output, the sentences have reduced repetition, but some of them are still incoherent:

```
The cat had no business entering the neighbors garage, but  it was too
late to stop it.
"I don't know if it was
The cat had no business entering the neighbors garage, but  she was
able to find her way to the porch, where she and her friend were
The cat had no business entering the neighbors garage, but  he did get
in the way.
The next day, the neighbor called the police
The cat had no business entering the neighbors garage, but  he managed
to get his hands on one of the keys, which he used to unlock
The cat had no business entering the neighbors garage, but  the
neighbors thought they were in the right place.
"What's going on
```

To improve the coherency, we can use another technique to include the next word from a set of words that have the highest likelihood of being the next word. We use the `top_k` parameter and set its value to 50. This instructs the generator to sample the next word from the top 50 words, arranged according to their probabilities.

We change the line in *step 4* to the following:

```
generated_sentences = generator(text, do_sample=True,
    max_length=30, num_return_sequences=5, num_beams=5,
    no_repeat_ngram_size=2, top_k=50, pad_token_id=50256)
```

```
The cat had no business entering the neighbors garage, but  it did get
into a neighbor's garage. The neighbor went to check on the cat
The cat had no business entering the neighbors garage, but  she was
there to take care of it.
```

```
The next morning, the cat was
```
```
The cat had no business entering the neighbors garage, but   it didn't
want to leave. The neighbor told the cat to get out of the
```
```
The cat had no business entering the neighbors garage, but   the
neighbors were too afraid to call 911.The neighbor told the police
that he
```
```
The cat had no business entering the neighbors garage, but   it was
there that he found his way to the kitchen, where it was discovered
that
```

We can also combine the `top_k` parameter with the `top_p` parameter. This instructs the generator to select the next word from the set of words that have a probability higher than this defined value. Adding this parameter with a value of `0.8` yields the following output:

```
generated_sentences = generator(text, do_sample=True,
    max_length=30, num_return_sequences=5, num_beams=5,
    no_repeat_ngram_size=2, top_k=50, top_p=0.8,
    pad_token_id=50256)
```

```
The cat had no business entering the neighbors garage, but   the owner
of the house told the police that he did not know what was going on
The cat had no business entering the neighbors garage, but   he did,
and the cat was able to get out of the garage.The
The cat had no business entering the neighbors garage, but   he was
able to get in through the back door. The cat was not injured,
The cat had no business entering the neighbors garage, but   the
neighbor told the police that the cat was a stray, and the neighbor
said that
The cat had no business entering the neighbors garage, but   the owner
of the house said he didn't know what to do with the cat.
```

As we can see, the addition of additional parameters to the generator continues to improve the generated output.

As a final example, let us generate a longer output sequence by changing the line in *step 4* to the following:

```
generated_sentences = generator(text, do_sample=True,
    max_length=500, num_return_sequences=1, num_beams=5,
    no_repeat_ngram_size=2, top_k=50, top_p=0.85,
    pad_token_id=50256)
```

```
The cat had no business entering the neighbors garage, but   she was
there to help.
"I was like, 'Oh my God, she's here,'" she said. "I'm like 'What are
you doing here?' "
The neighbor, who asked not to be identified, said she didn't know
what to make of the cat's behavior. She said it seemed like it was
trying to get into her home, and that she was afraid for her life. The
```

```
neighbor said that when she went to check on her cat, it ran into the
neighbor's garage and hit her in the face, knocking her to the ground.
```

As we can see, the generated output, however fictitious, is more coherent and readable. We encourage you to experiment with different mixes of parameters and their respective values to improve the generated output based on their use cases.

Please note that the output returned by the model might differ a bit from what this example has shown. This happens because the internal language model is probabilistic in nature. The next word is sampled from a distribution that contains words that have a probability larger than what we defined in our parameters for generation.

In this recipe, we used the decoder module of the transformer to generate text, given a seed sentence. There are use cases where an encoder and decoder are used together to generate text. We will learn about this in the next recipe.

Language translation

In this recipe, we will use transformers for language translation. We will use the **Google Text-To-Text Transfer Transformer** (**T5**) model. This model is an end-to-end model that uses both the encoder and decoder components of the transformer model.

Getting ready

As part of this recipe, we will use the pipeline module from the transformers package. You can use the 8.6_Language_Translation_with_transformers.ipynb notebook from the code site if you need to work from an existing notebook.

How to do it...

In this recipe, you will initialize a seed sentence in English and translate it to French. The T5 model expects the input format to encode the information about the language translation task along with the seed sentence. In this case, the encoder uses the input in the source language and generates a representation of the text. The decoder uses this representation and generates text for the target language. The T5 model is trained specifically for this task, in addition to many others. If you are running on a machine that does not have a CUDA-compatible device, it might take some time for the recipe steps to be executed.

The recipe does the following things:

- It initializes the Google t5-base model and tokenizer
- It initializes a seed sentence in English that will be translated into French

- It tokenizes the seed sentence along with the task specification to translate the seed sentence into French

- It generates the translated tokens, decodes them into the target language (French), and prints them

The steps for the recipe are as follows:

1. Do the necessary imports and identify the compute device, as described in the previous classification recipe:

```
from transformers import (
    T5Tokenizer, T5ForConditionalGeneration)
import torch
device = torch.device("cuda" if torch.cuda.is_available() else
"cpu")
```

2. Initialize a tokenizer and model instance with the t5-base model from Google. We use the model_max_length parameter of 200. Feel free to experiment with higher values if your seed sentence is longer than 200 words. We also load the model onto the device that was identified for computation in step 1:

```
tokenizer = T5Tokenizer.from_pretrained(
    't5-base', model_max_length=200)
model = T5ForConditionalGeneration.from_pretrained(
    't5-base', return_dict=True)
model = model.to(device)
```

3. Initialize a seed sequence that you want to translate:

```
language_sequence = ("It's such a beautiful morning today!")
```

4. Tokenize the input sequence. The tokenizer specifies the source and the target language as part of its input encoding. This is done by appending the "translate English to French:" text to the input seed sequence. We load these token IDs into the device that is used for computation. It is a requirement for both the model and the token IDs to be on the same device:

```
input_ids = tokenizer(
    "translate English to French: " + language_sequence,
    return_tensors="pt",
    truncation=True).input_ids.to(device)
```

5. Translate the source language token IDs to the target language token IDs via the model. The model uses the encoder-decoder architecture to convert the input token IDs to the output token IDs:

```
language_ids = model.generate(input_ids, max_new_tokens=200)
```

6. Decode the text from the token IDs to the target language tokens. We use the tokenizer to convert the output token IDs to the target language tokens:

```
language_translation = tokenizer.decode(
    language_ids[0], skip_special_tokens=True)
```

7. Print the translated output:

```
print(language_translation)
C'est un matin si beau!
```

In conclusion, this chapter introduced the concept of transformers, along with some of its basic applications. The next chapter will focus on how we can use the different NLP techniques to understand text better.

9
Natural Language Understanding

In this chapter, we will explore recipes that will allow us to interpret and understand the text contained in short as well as long passages. **Natural language understanding** (**NLU**) is a very broad term and the various systems developed as part of NLU do not interpret or understand a passage of text the same way a human reader would. However, based on the specificity of the task, we can create some applications that can be combined to generate an interpretation or understanding that can be used to solve a given problem related to text processing.

Organizations that have a huge document corpus need a seamless way to search through documents. More specifically, what users really need is an answer to a specific question without having to glean through a list of documents that are returned as part of a document search. Users would prefer the query to be formulated as a question in natural language and the answer to be emitted in the same manner.

Another set of applications is that of document summarization and text entailment. While processing a large set of documents, it is helpful if the document length can be shortened without the loss of meaning or context. Additionally, it's important to determine whether the information contained in the document at the sentence level entails itself.

While we work on processing and classifying the documents, there are always challenges in understanding why or how the model assigns a label to a piece of text – more specifically, what parts of the text contribute to the different labels.

This chapter will cover different techniques to explore the various aspects previously described. We will follow recipes that will allow us to perform these tasks and understand the underlying building blocks that help us achieve the end goals.

As part of this chapter, we will build recipes for the following tasks:

- Answering questions from a short text passage
- Answering questions from a long text passage

- Answering questions from a document corpus in an extractive manner
- Answering questions from a document corpus in an abstractive manner
- Summarizing text using pretrained models based on Transformers
- Detecting sentence entailment
- Enhancing explainability via a classifier-invariant approach
- Enhancing explainability via text generation

Technical requirements

The code for this chapter is in a folder named `Chapter9` in the GitHub repository of the book (`https://github.com/PacktPublishing/Python-Natural-Language-Processing-Cookbook-Second-Edition/tree/main/Chapter09`).

As in previous chapters, the packages required for this chapter are part of the `poetry` environment. Alternatively, you can install all the packages using the `requirements.txt` file.

Answering questions from a short text passage

To get started with question answering, we will start with a simple recipe that can answer a question from a short passage.

Getting ready

As part of this chapter, we will use the libraries from Hugging Face (`huggingface.co`). For this recipe, we will use the `BertForQuestionAnswering` and `BertTokenizer` modules from the Transformers package. The `BertForQuestionAnswering` model uses the base BERT large uncased model that was trained on the SQuAD dataset and fine-tuned for the question-answering task. This pre-trained model can be used to load a text passage and answer questions based on the contents of the passage. You can use the `9.1_question_answering.ipynb` notebook from the code site if you need to work from an existing notebook.

How to do it...

In this recipe, we will load a pretrained model that has been trained on the SQuAD dataset (`https://huggingface.co/datasets/squad`).

The recipe does the following things:

- It initializes a question-answering pipeline based on the pre-trained `BertForQuestionAnswering` model and `BertTokenizer` tokenizer.

- It further initializes a context passage and a question and emits the output of the answer based on these two parameters. It also prints the exact text of the answer.

- It asks a follow-up question to the same pipeline by just changing the question text, and prints the exact text answer to the question.

The steps for the recipe are as follows:

1. Do the necessary imports to import the necessary types and functions from the datasets package:

```
from transformers import (
    pipeline, BertForQuestionAnswering, BertTokenizer)
import torch
```

2. In this step, we initialize the model and tokenizer, respectively, using the pre-trained bert-large-uncased-whole-word-masking-finetuned-squad artifacts. These will be downloaded from the Hugging Face website if they are not present locally on the machine as part of these calls. We have chosen the specific model and tokenizer for our recipe, but feel free to explore other models on the Hugging Face site that might suit your needs. As a generic step for this and the following recipe, we discover whether there are any GPU devices in the system and attempt to use them. If a GPU is not detected, we use the CPU instead:

```
device = torch.device("cuda" if torch.cuda.is_available()
    else "cpu")
qa_model = BertForQuestionAnswering.from_pretrained(
    'bert-large-uncased-whole-word-masking-finetuned-squad',
    device_map=device)
qa_tokenizer = BertTokenizer.from_pretrained(
    'bert-large-uncased-whole-word-masking-finetuned-squad',
    device=device)
```

3. In this step, we initialize a question-answering pipeline with the model and tokenizer. The task type for this pipeline is set to question-answering:

```
question_answer_pipeline = pipeline(
    "question-answering", model=qa_model,
    tokenizer=qa_tokenizer)
```

4. In this step, we initialize a context passage. This passage was generated as part of our *Text Generation via Transformers* example in *Chapter 8*. It's entirely acceptable if you want to use a different passage:

```
context = "The cat had no business entering the neighbors
garage, but she was there to help. The neighbor, who asked not
to be identified, said she didn't know what to make of the cat's
behavior. She said it seemed like it was trying to get into
her home, and that she was afraid for her life. The neighbor
```

said that when she went to check on her cat, it ran into the
neighbor's garage and hit her in the face, knocking her to the
ground."

5. In this step, we initialize a question text, invoke the pipeline with the context and question, and store the result in a variable. The type of the result is a Python `dict` object:

```
question = "Where was the cat trying to enter?"
result = question_answer_pipeline(question=question,
    context=context)
```

6. In this step, we print the value of the result. The `score` value shows the probability of the answer. The `start` and `end` values denote the start and end character indices in the `context` passage that constitute the answer. The `answer` value denotes the actual text of the answer:

```
print(result)
```

```
{'score': 0.25, 'start': 33, 'end': 54, 'answer': 'the neighbors
garage,'}
```

7. In this step, we print the exact text answer. This is present in the `answer` key in the `result` dictionary:

```
print(result['answer'])
```

```
the neighbors garage,
```

8. In this step, we ask another question using the same context and print the result:

```
question = "What did the cat do after entering the garage"
result = question_answer_pipeline(
    question=question, context=context)
print(result['answer'])
```

```
hit her in the face, knocking her to the ground.
```

Answering questions from a long text passage

In the previous recipe, we learned an approach to extract the answer to a question, given a context. This pattern involves the model retrieving the answer from the given context. The model cannot answer a question that is not contained in the context. This does serve a purpose where we want an answer from a given context. This type of question-answering system is defined as **Closed Domain Question Answering (CDQA)**.

There is another system of question answering that can answer questions that are general in nature. These systems are trained on larger corpora. This training provides them with the ability to answer questions that are open in nature. These systems are called **Open Domain Question Answering (ODQA)** systems.

Getting ready

As part of this recipe, we will use the **DeepPavlov** (https://deeppavlov.ai) ODQA system to answer an open question. We will use the deeppavlov library along with the **Knowledge Base Question Answering** (KBQA) model. This model has been trained on English wiki data as a knowledge base. It uses various NLP techniques such as entity linking and disambiguation, knowledge graphs, and so on to extract the exact answer to the question.

This recipe needs a few steps to set up the right environment for its execution. The poetry file for this recipe is in the 9.2_QA_on_long_passages folder. We will also need to install and download the document corpus by performing the following command:

```
python -m deeppavlov install kbqa_cq_en
```

You can also use the 9.2_QA_on_long_passages.ipynb notebook, which is contained within the same folder.

How to do it...

In this recipe, we will initialize the KBQA model based on the DeepPavlov library and use it to answer an open question. The steps for the recipe are as follows:

1. Do the necessary imports:

    ```
    from deeppavlov import build_model
    ```

2. In this step, we initialize the KBQA model, kbqa_cq_en, which is passed to the build_model method as an argument. We also set the download argument to True so that the model is downloaded as well in case it is missing locally:

    ```
    kbqa_model = build_model('kbqa_cq_en', download=True)
    ```

3. We use the initialized model and pass it a couple of questions that we want to be answered:

    ```
    result = kbqa_model(['What is the capital of Egypt?',
        'Who is Bill Clinton\'s wife?'])
    ```

4. We print the result as returned by the model. The result contains three arrays.

    The first array contains the exact answers to the question ordered in the same way as the original input. In this case, the answers Cairo and Hillary Clinton are in the same order as the questions they pertain to.

    You might observe some additional artifacts in the output. These are internal identifiers that are generated by the library. We have omitted them for brevity:

    ```
    [['Cairo', 'Hillary Clinton']]
    ```

See also

For more information on the internal details of the working of DeepPavlov, please refer to `https://deeppavlov.ai`.

Answering questions from a document corpus in an extractive manner

For the use cases where we have a document corpus that contains a large number of documents, it's not feasible to load the document content at runtime to answer a question. Such an approach would lead to long query times and would not be suitable for production-grade systems.

In this recipe, we will learn how to preprocess the documents and transform them into a form for faster reading, indexing, and retrieval that allows the system to extract the answer for a given question with short query times.

Getting ready

As part of this recipe, we will use the **Haystack** (`https://haystack.deepset.ai/`) framework to build a **QA system** that can answer questions from a document corpus. We will download a dataset based on *Game of Thrones* and index it. For our QA system to be performant, we will need to index the documents beforehand. Once the documents are indexed, answering a question follows a two-step process:

1. **Retriever**: Since we have many documents, scanning each document to fetch an answer is not a feasible approach. We will first retrieve a set of candidate documents that can possibly contain an answer to our question. This step is performed using a `Retriever` component. This searches through the pre-created index to filter the number of documents that we will need to scan to retrieve the exact answer.

2. **Reader**: Once we have a candidate set of documents that could contain the answer, we will search these documents to retrieve the exact answer to our question.

We will discuss the details of these components throughout this recipe. You can use the `9.3_QA_on_document_corpus.ipynb` notebook from the code site if you need to work from an existing notebook. To start with, let's set up the prerequisites.

How to do it...

1. In this step, we do the necessary imports:

```
import os
from haystack.document_stores import InMemoryDocumentStore
from haystack.nodes import BM25Retriever, FARMReader
```

```
from haystack.pipelines import ExtractiveQAPipeline
from haystack.pipelines.standard_pipelines import(
    TextIndexingPipeline)
from haystack.utils import (fetch_archive_from_http,
    print_answers)
```

2. In this step, we specify a folder that will be used to save our dataset. Then, we retrieve the dataset from the source. The second parameter to the `fetch_archive_from_http` method is the folder in which the dataset will be downloaded. We set the parameter to the folder that we defined in the first line. The `fetch_archive_from_http` method decompresses the archive `.zip` file and extracts all files into the same folder. We then read from the folder and create a list of files contained in the folder. We also print the number of files that are present:

```
doc_dir = "data/got_dataset"
fetch_archive_from_http(
    url="https://s3.eu-central-1.amazonaws.com/deepset.
ai-farm-qa/datasets/documents/wiki_gameofthrones_txt1.zip",
    output_dir=doc_dir,
    )
files_to_index = [doc_dir + "/" + f for f in os.listdir(
    doc_dir)]
print(len(files_to_index))
183
```

3. We initialize a document store based on the files. We create an indexing pipeline based on the document store and execute the indexing operation. To achieve this, we initialize an `InMemoryDocumentStore` instance. In this method call, we set the `use_bm25` argument as `True`. The document store uses **Best Match 25** (**bm25**) as the algorithm for the retriever step. The `bm25` algorithm is a simple bag-of-words-based algorithm that uses a scoring function. This function utilizes the number of times a term is present in the document and the length of the document. *Chapter 3* covers the `bm25` algorithm in more detail and we recommend you refer to that chapter for better understanding. Note that there are various other `DocumentStore` options such as `ElasticSearch`, `OpenSearch`, and so on. We used an `InMemoryDocumentStore` document store to keep the recipe simple and focus on the retriever and reader concepts:

```
document_store = InMemoryDocumentStore(use_bm25=True)
indexing_pipeline = TextIndexingPipeline(document_store)
indexing_pipeline.run_batch(file_paths=files_to_index)
```

4. Once we have loaded the documents, we initialize our retriever and reader instances. To achieve this, we initialize the retriever and the reader components. `BM25Retriever` uses the `bm25` scoring function to retrieve the initial set of documents. For the reader, we initialize the `FARMReader` object. This is based on deepset's FARM framework, which can utilize the QA

models from Hugging Face. In our case, we use the `deepset/roberta-base-squad2` model as a reader. The `use_gpu` argument can be set appropriately based on whether your device has a GPU or not:

```
retriever = BM25Retriever(document_store=document_store)
reader = FARMReader(
    model_name_or_path="deepset/roberta-base-squad2",
    use_gpu=True)
```

5. We now create a pipeline that we can use to answer questions. After having initialized the retriever and reader in the previous step, we want to combine them for querying. The `pipeline` abstraction from the Haystack framework allows us to integrate the reader and retriever together using a series of pipelines that address different use cases. In this instance, we will use `ExtractiveQAPipeline` for our QA system. After the initialization of the pipeline, we generate the answer to a question from the *Game of Thrones* series. The `run` method takes the question as the query. The second argument, `params`, dictates how the results from the retriever and reader are combined to present the answer:

    - `"Retriever": {"top_k": 10}`: The `top_k` keyword argument specifies that the top-k (in this case, 10) results from the retriever are used by the reader to search for the exact answer

    - `"Reader": {"top_k": 5}`: The `top_k` keyword argument specifies that the top-k (in this case, 5) results from the reader are presented as the output of the method:

```
pipe = ExtractiveQAPipeline(reader, retriever)
prediction = pipe.run(
    query="Who is the father of Arya Stark?",
    params={"Retriever": {"top_k": 10}, "Reader": {"top_k": 5}}
)
```

6. We print the answer to our question. The system prints out the exact answer along with the associated context that it used to extract the answer from. Note that we use the value of `all` for the `details` argument. Using the `all` value for the same argument prints out `start` and `end` spans for the answer along with all the auxiliary information. Setting the value of `medium` for the `details` argument provides the relative score of each answer. This score can be used to filter out the results further based on the accuracy requirements of the system. Using the argument of `medium` presents only the answer and the context. We encourage you to make a suitable choice based on your requirements:

```
print_answers(prediction, details="all")
'Query: Who is the father of Arya Stark?'
'Answers:'
[<Answer {'answer': 'Eddard',
'type': 'extractive',
'score': 0.993372917175293,
```

'context': "s Nymeria after a legendary warrior queen. She
travels with her father, Eddard, to King's Landing when he
is made Hand of the King. Before she leaves,", 'offsets_
in_document': [{'start': 207, 'end': 213}], 'offsets_
in_context': [{'start': 72, 'end': 78}], 'document_ids':
['9e3c863097d66aeed9992e0b6bf1f2f4'], 'meta': {'_split_id':
3}}>,

<Answer {'answer': 'Ned',
'type': 'extractive',
'score': 0.9753613471984863,
'context': "k in the television series.\n\n====Season 1====\
nArya accompanies her father Ned and her sister Sansa to
King's Landing. Before their departure, Arya's h", 'offsets_
in_document': [{'start': 630, 'end': 633}], 'offsets_in_
context': [{'start': 74, 'end': 77}], 'document_ids':
['7d3360fa29130e69ea6b2ba5c5a8f9c8'], 'meta': {'_split_id':
10}}>,

<Answer {'answer': 'Lord Eddard Stark',
'type': 'extractive',
'score': 0.9177322387695312,
'context': 'rk daughters.\n\nDuring the Tourney of the
Hand to honour her father Lord Eddard Stark, Sansa Stark
is enchanted by the knights performing in the event.',
'offsets_in_document': [{'start': 280, 'end': 297}], 'offsets_
in_context': [{'start': 67, 'end': 84}], 'document_ids':
['5dbccad397381605eba063f71dd500a6'], 'meta': {'_split_id':
3}}>,

<Answer {'answer': 'Ned',
'type': 'extractive',
'score': 0.8396496772766113,
'context': " girl disguised as a boy all along and is
surprised to learn she is Arya, Ned Stark's daughter. After
the Goldcloaks get help from Ser Amory Lorch and", 'offsets_
in_document': [{'start': 848, 'end': 851}], 'offsets_in_
context': [{'start': 74, 'end': 77}], 'document_ids':
['257088f56d2faba55e2ef2ebd19502dc'], 'meta': {'_split_id':
31}}>,

<Answer {'answer': 'King Robert',
'type': 'extractive',
'score': 0.6922298073768616,
'context': "en refuses to yield Gendry, who is actually a
bastard son of the late King Robert, to the Lannisters. The
Night's Watch convoy is overrun and massacr", 'offsets_in_
document': [{'start': 579, 'end': 590}], 'offsets_in_context':
[{'start': 70, 'end': 81}], 'document_ids': ['4d51b1876e8a7eac81
32b97e2af04401'], 'meta': {'_split_id': 4}}>]

See also

For a QA system to work in a high-performance production system, it is recommended to use a different document store from an in-memory one. We recommend you refer to `https://docs.haystack.deepset.ai/docs/document_store` and use an appropriate document store based on your production-grade requirements.

Answering questions from a document corpus in an abstractive manner

In the previous recipe, we learned how to build a QA system based on the document corpora. The answers that were retrieved were extractive in nature (i.e., the answer snippet was a piece of text copied verbatim from the document source). There are techniques to generate an abstractive answer too, which is more readable by end users compared to an extractive one.

Getting ready

For this recipe, we will build a QA system that will provide answers that are abstractive in nature. We will load the `bilgeyucel/seven-wonders` dataset from the Hugging Face site and initialize a retriever from it. This dataset has content about the seven wonders of the ancient world. To generate the answers, we will use the `PromptNode` component from the Haystack framework to set up a pipeline that can generate answers in an abstractive fashion. You can use the `9.4_abstractive_qa_on_document_corpus.ipynb` notebook from the code site if you need to work from an existing notebook. Let's get started.

How to do it

The steps are as follows:

1. In this step, we do the necessary imports:

```
from datasets import load_dataset
from haystack.document_stores import InMemoryDocumentStore
from haystack.nodes import (
    BM25Retriever, PromptNode,
    PromptTemplate, AnswerParser)
from haystack.pipelines import Pipeline
```

2. As part of this step, we load the `bilgeyucel/seven-wonders` dataset into an in-memory document store. This dataset has been created out of the Wikipedia pages of *Seven Wonders of the Ancient World* (`https://en.wikipedia.org/wiki/Wonders_of_the_World`). This dataset has been preprocessed and uploaded to the Hugging Face site, and can be easily downloaded by using the `datasets` module from Hugging Face. We use

`InMemoryDocumentStore` as our document store, with `bm25` as the search algorithm. We write the documents from the dataset into the document store. To have a performant query time performance, the `write_documents` method automatically optimizes how the documents are written. Once the documents are written into, we initialize the retriever based on `bm25`, similar to our previous recipe:

```
dataset = load_dataset("bilgeyucel/seven-wonders",
    split="train")
document_store = InMemoryDocumentStore(use_bm25=True)
document_store.write_documents(dataset)
retriever = BM25Retriever(document_store=document_store)
```

3. As part of this step, we initialize a prompt template. We can define the task we want the model to perform as a simple instruction in English using the `prompt` argument. It also takes two inline arguments, `document` and `query`. These arguments are expected to be in the execution context at runtime. The second argument, `output_parser`, takes an `AnswerParser` object. This object instructs the `PromptNode` object to store the results in the `answers` element. After defining the prompt, we initialize a `PromptNode` object with a model and the prompt template. We use the `google/flan-t5-large` model as the answer generator. This model is based on the Google T5 language model and has been fine-tuned (`flan` stands for **fine-tuning language models**). Fine-tuning a language model with an instruction dataset allows the language model to perform tasks following simple instructions and generating text based on the given context and instruction. One of the fine-tuning steps as part of this model training was to operate on human written instructions as tasks. This allowed the model to perform different downstream tasks on instructions alone and reduced the need for any few-shot examples to be trained on.

```
rag_prompt = PromptTemplate(
    prompt="""Synthesize a comprehensive answer from the
following text for the given question.
        Provide a clear and concise response that summarizes the
key points and information presented in the text.
        Your answer should be in your own words and be no longer
than 50 words.
        \n\n Related text: {join(documents)} \n\n Question:
{query} \n\n Answer:""",
    output_parser=AnswerParser(),
)
prompt_node = PromptNode(
    model_name_or_path="google/flan-t5-large",
    default_prompt_template=rag_prompt, use_gpu=True)
```

4. We now create a pipeline and add the `retriever` and `prompt_node` components that we initialized in the previous steps. The `retriever` component operates on the query supplied by the user and generates a set of results. These results are passed to the prompt node, which uses the configured `flan-t5-model` to generate the answer:

```
pipe = Pipeline()
pipe.add_node(component=retriever, name="retriever",
    inputs=["Query"])
pipe.add_node(component=prompt_node,
    name="prompt_node", inputs=["retriever"])
```

5. Once the pipeline is set up, we use it to answer questions on the content based on the dataset:

```
output = pipe.run(query="What is the Great Pyramid of Giza?")
print(output["answers"][0].answer)
output = pipe.run(query="Where are the hanging gardens?")
print(output["answers"][0].answer)
```

```
The Great Pyramid of Giza was built in the early 26th century BC
during a period of around 27 years.[3]
The Hanging Gardens of Semiramis are the only one of the Seven Wonders
for which the location has not been definitively established.
```

See also

Please refer to the prompt engineering guide on Haystack on how to generate prompts for your use cases (`https://docs.haystack.deepset.ai/docs/prompt-engineering-guidelines`).

Summarizing text using pre-trained models based on Transformers

We will now explore techniques for performing text summarization. Generating a summary for a long passage of text allows NLP practitioners to extract the relevant information for their use cases and use these summaries for other downstream tasks. As part of the summarization, we will explore recipes that use Transformer models to generate the summaries.

Getting ready

Our first recipe for summarization will use the Google **Text-to-Text Transfer Transformer** (T5) model for summarization. You can use the `9.5_summarization.ipynb` notebook from the code site if you need to work from an existing notebook.

How to do it

Let's get started:

1. Do the necessary imports:

    ```
    from transformers import pipeline
    ```

2. As part of this step, we initialize the input passage that we need to summarize along with the pipeline. We also calculate the length of the passage since this will be used as an argument to be passed to the pipeline during the task execution in the next step. Since we have defined the task as `summarization`, the object returned by the pipeline module is of the `SummarizationPipeline` type. We also pass `t5-large` as the model parameter for the pipeline. This model is based on the `Encoder-Decoder` Transformer model and acts as a pure sequence-to-sequence model. That means the input and output to/from the model are text sequences. This model was pre-trained using the denoising objective of finding masked words in a sentence followed by fine-tuning on specific downstream tasks such as summarization, textual entailment, language translation, and so on:

    ```
    passage = "The color of animals is by no means a matter of
    chance; it depends on many considerations, but in the majority
    of cases tends to protect the animal from danger by rendering
    it less conspicuous. Perhaps it may be said that if coloring is
    mainly protective, there ought to be but few brightly colored
    animals. There are, however, not a few cases in which vivid
    colors are themselves protective. The kingfisher itself, though
    so brightly colored, is by no means easy to see. The blue
    harmonizes with the water, and the bird as it darts along the
    stream looks almost like a flash of sunlight."

    passage_length = len(passage.split(' '))
    pipeline_instance = pipeline("summarization", model="t5-large")
    ```

3. We now use the `pipeline_instance` initialized in the previous step and pass the text passage to it to perform the `summarization` step. A string array can be passed as well if multiple sequences are to be summarized. We pass `max_length=512` as the second argument. The T5 model is memory-intensive and the compute requirements grow quadratically with the increase in the input text length. This step might take a few minutes to complete based on the compute capability of the environment you are executing this on:

    ```
    pipeline_result = pipeline_instance(
        passage, max_length=passage_length)
    ```

4. Once the `summarization` step is complete, we extract the result from the output and print it. The pipeline returns a list of dictionaries. Each list item corresponds to the input argument. In this case, since we passed only one string as input, the first item in the list is the output

dictionary that contains our summary. The summary can be retrieved by indexing the dictionary on the `summary_text` element:

```
result = pipeline_result[0]["summary_text"]
print(result)
```

```
the color of animals is by no means a matter of chance; it depends
on many considerations . in the majority of cases, coloring tends to
protect the animal from danger . there are, however, not a few cases
in which vivid colors are themselves protective .
```

There's more...

Now that we have seen how we can generate a summary using the T5 model, we can use the same code framework and tweak it slightly to use other models to generate summaries.

The following lines would be common for the other summarization recipes that we are using. We added an extra variable named `device`, which we will use in our pipelines. We set this variable to the value of the device that we will use to generate the summary. If a GPU is present and configured in the system, it will be used; otherwise, the summarization will be performed using the CPU:

```
from transformers import pipeline
import torch
device = torch.device("cuda" if torch.cuda.is_available() else "cpu")
passage = "The color of animals is by no means a matter of chance; it
depends on many considerations, but in the majority of cases tends
to protect the animal from danger by rendering it less conspicuous.
Perhaps it may be said that if coloring is mainly protective, there
ought to be but few brightly colored animals. There are, however,
not a few cases in which vivid colors are themselves protective. The
kingfisher itself, though so brightly colored, is by no means easy
to see. The blue harmonizes with the water, and the bird as it darts
along the stream looks almost like a flash of sunlight."
```

In the following example, we use the **BART** model (`https://huggingface.co/facebook/bart-large-cnn`) from Facebook. This model was trained using a denoising objective. A function adds some random piece of text to an input sequence. The model is trained based on the objective to denoise or remove the noisy text from the input sequence. The model was further fine-tuned using the **CNN DailyMail** dataset (`https://huggingface.co/datasets/abisee/cnn_dailymail`) for summarization:

```
pipeline_instance = pipeline("summarization",
    model="facebook/bart-large-cnn", device=device)
pipeline_result = pipeline_instance(passage,
    max_length=passage_length)
result = pipeline_result[0]["summary_text"]
print(result)
```

```
The color of animals is by no means a matter of chance; it depends on
many considerations, but in the majority of cases tends to protect
the animal from danger by rendering it less conspicuous. There
are, however, not a few cases in which vivid colors are themselves
protective. The blue harmonizes with the water, and the bird as it
darts along the stream looks almost like a flash of sunlight.
```

As we observe from the generated summary, it is verbose and extractive in nature. Let's try generating a summary with another model.

In the following example, we use the **PEGASUS** model from Google (`https://huggingface.co/google/pegasus-large`) for summarization. This model is a Transformer-based Encoder-Decoder model that was pre-trained with a large news and web page corpus – C4 (`https://huggingface.co/datasets/allenai/c4`) and the HugeNews dataset – on a training objective of detecting important sentences. HugeNews is a dataset of 1.5 billion articles curated from news and news-like websites from 2013–2019. This model was further fine-tuned for summarization using the subset of the same dataset. The training objective for the fine-tuning involved masking important sentences and making the model generate a summary that has these important sentences. This model generates abstract summaries:

```
pipeline_instance = pipeline("summarization",
    model="google/pegasus-large", device=device)
pipeline_result = pipeline_instance([passage, passage],
    max_length=passage_length)
result = pipeline_result[0]["summary_text"]
print(result)
```

```
Perhaps it may be said that if coloring is mainly protective, there
ought to be but few brightly colored animals.
```

As we observe from the generated summary, it is concise and abstractive.

See also

As many new and improved models for summarization are always in the works, we recommend that you refer to the models on the Hugging Face site (`https://huggingface.co/models?pipeline_tag=summarization`) and make the respective choice based on your requirements.

Detecting sentence entailment

In this recipe, we will explore techniques to detect **textual entailment**, given a set of two sentences. The first sentence in the set is the `premise`, which sets up a context. The second sentence is the `hypothesis`, which serves as the claim. Textual entailment identifies the contextual relationship between the `premise` and the `hypothesis`. These relationships can be of three types, defined as follows:

- **Entailment** – The hypothesis supports the premise
- **Contradiction** – The hypothesis contradicts the premise
- **Neutral** – The hypothesis neither supports nor contradicts the premise

Getting ready

We will use the Transformers library to detect text entailment. You can use the `9.6_textual_entailment.ipynb` notebook from the code site if you need to work from an existing notebook.

How to do it...

In this recipe, we will initialize different sets of sentences that are related through each of the previously defined relationships and explore methods to detect these relationships. Let's get started:

1. Do the necessary imports:

    ```
    import torch
    from transformers import T5Tokenizer, T5ForConditionalGeneration
    ```

2. Initialize the device, the tokenizer, and the model. In this case, we are using Google's `t5-small` model. We set the `legacy` flag to `False` since we don't need to use the legacy behavior of the model. We set the `device` value based on whatever device we have available in our execution environment. Similarly, for the model, we set the `model` name and `device` parameter similar to the tokenizer. We set the `return_dict` parameter as `True` so that we get the model results as a dictionary instead of a tuple:

    ```
    device = torch.device("cuda" if torch.cuda.is_available()
        else "cpu")
    tokenizer = T5Tokenizer.from_pretrained(
        't5-small', legacy=False, device=device)
    model = T5ForConditionalGeneration.from_pretrained(
        't5-small', return_dict=True, device_map=device)
    ```

3. We initialize the `premise` and `hypothesis` sentences. In this case, the hypothesis supports the premise:

    ```
    premise = "The corner coffee shop serves the most awesome coffee
    I have ever had."
    hypothesis = "I love the coffee served by the corner coffee
    shop."
    ```

4. In this step, we call the tokenizer with the `mnli premise` and `hypothesis` values. This is a simple text concatenation step to set up the tokenizer for the `entailment` task. We read the `input_ids` property to get the token identifiers for the concatenated string. Once we have the token IDs, we use the model to generate the entailment prediction. This returns a list of tensors with the predictions, which we use in the next step:

```
input_ids = tokenizer(
    "mnli premise: " + premise + " hypothesis: " + hypothesis,
    return_tensors="pt").input_ids
entailment_ids = model.generate(input_ids.to(device),
    max_new_tokens=20)
```

5. In this step, we call the `decode` method of the tokenizer and pass it the first tensor (or vector) of the tensors that were returned by the `generate` call of the model. We also instruct the tokenizer to skip the special tokens that are used by the tokenizer internally. The tokenizer generates the string label from the vector that is passed in. We print the prediction result. In this case, the generated prediction by the model is `entailment`:

```
prediction = tokenizer.decode(
    entailment_ids[0], skip_special_tokens=True, device=device)
print(prediction)
```

entailment

There's more...

Now that we have shown an example in the case of entailment with a single sentence, the same framework can be used to process a batch of sentences to generate entailment predictions. We will tailor *steps 3,4*, and *5* from the previous recipe for this example. We initialize an array of two sentences for both `premise` and `hypothesis`, respectively. Both the `premise` sentences are the same, while the `hypothesis` sentences are of `entailment` and `contradiction`, respectively:

```
premise = ["The corner coffee shop serves the most awesome coffee
I have ever had.", "The corner coffee shop serves the most awesome
coffee I have ever had."]
hypothesis = ["I love the coffee served by the corner coffee shop.",
"I find the coffee served by the corner coffee shop too bitter for my
taste."]
```

Since we have an array of sentences for both `premises` and `hypothesis`, we create an array of concatenated inputs that combine the `tokenizer` instruction. This array is used to pass to the tokenizer and we use the token IDs returned by `tokenizer` in the next step:

```
premises_and_hypotheses = [f"mnli premise: {pre}
    hypothesis: {hyp}" for pre, hyp in zip(premise, hypothesis)]
input_ids = tokenizer(
```

```
text=premises_and_hypotheses, padding=True,
return_tensors="pt").input_ids
```

We now generate the predictions using the same methodology that we used earlier. However, in this step, we generate the inference label by iterating through the tensors returned by the model output and printing the prediction:

```
entailment_ids = model.generate(input_ids.to(device),
    max_new_tokens=20)
for _tensor in entailment_ids:
    entailment = tokenizer.decode(_tensor,
        skip_special_tokens=True, device=device)
    print(entailment)
```

Enhancing explainability via a classifier-invariant approach

Now, we will explore recipes that will allow us to understand the decisions made by text classifiers. We will explore techniques that will use a sentiment classifier and NLP explainability libraries to interpret the classification labels and their relation to the input text, especially in the aspect of individual words in the text.

Though a lot of the current models for text classification in NLP are based on deep neural networks, it is difficult to interpret the results of classification via the network weights or parameters. It is equally challenging to map these network parameters to the individual components or words in the input. However, there are still a few techniques in the NLP space to help us understand the decisions made by the classifier. We will explore these techniques in the current recipe and the following one.

In this recipe, we will learn how to interpret the feature importance of each word in a text passage while being invariant of the classifier model. This technique can be used for any text classifier as we treat the classifier as a black box and use the results of the predictions to infer the results from an explainability perspective.

Getting ready

We will use the lime library for explainability. You can use the 9.7_explanability_via_classifier.ipynb notebook from the code site if you want to work from an existing notebook.

How to do it...

In this recipe, we will repurpose a classifier that we built in the *Transformers* chapter and use it to generate a sentiment prediction. We will call this classifier multiple times with a perturbation of the input to understand the contribution of each word to the sentiment. Let's get started:

1. Do the necessary imports:

    ```
    import numpy as np
    import torch

    from lime.lime_text import LimeTextExplainer
    from transformers import pipeline
    ```

2. In this step, we initialize the device and the pipeline for sentiment classification. For more details on this step, please refer to chapter-8.

    ```
    device = torch.device(
        "cuda" if torch.cuda. is_available() else "cpu")
    roberta_pipe = pipeline(
        "sentiment-analysis",
        model="siebert/sentiment-roberta-large-english",
        tokenizer="siebert/sentiment-roberta-large-english",
        top_k=1,
        device=device
    )
    ```

3. In this step, we initialize a sample text passage along with setting the print options. Setting the print options allows us to print the outputs in the later steps in an easy-to-read format:

    ```
    sample_text = "I really liked the Oppenheimer movie and found it
    truly entertaining and full of substance."
    np.set_printoptions(suppress = True,
        formatter = {'float_kind':'{:f}'.format},
        precision = 2)
    ```

4. In this step, we create a wrapper function for sentiment classification. This method is used by the explainer to invoke the classification pipeline multiple times to gauge the contribution of each word in the passage:

    ```
    def predict_prob(texts):
        preds = roberta_pipe(texts)
        preds = np.array([
            [label[0]['score'], 1 - label[0]['score']]
            if label[0]['label'] == 'NEGATIVE'
            else [1 - label[0]['score'], label[0]['score']]
    ```

```
        for label in preds
    ])
    return preds
```

5. In this step, we instantiate the `LimeTextExplainer` class and call the `explain_instance` method for it. This method takes the sample text along with the `classifier` wrapper function. The wrapper function passed to this method expects it to take a single instance of a string and return the probabilities of the target classes. In this case, our wrapper function accepts a simple string and returns the probabilities for the `NEGATIVE` and `POSITIVE` classes, respectively, and in that order:

```
explainer = LimeTextExplainer(
    class_names=['NEGATIVE', 'POSITIVE'])
exp = explainer.explain_instance(
    text_instance=sample_text,
    classifier_fn=predict_prob)
```

6. In this step, we print the class probabilities for the sample text. As we observe, the sample text has been assigned a `POSITIVE` sentiment as per the classifier:

```
original_prediction = predict_prob(sample_text)
print(original_prediction)
```

```
[[0.001083 0.998917]]
```

7. In this step, we print the explanations. As we observe from the probabilities for each word, the words `entertaining` and `liked` contributed the most to the `POSITIVE` class. There are some words that contribute negatively to the positive sentiment, but overall, the sentence is classified as positive:

```
print(np.array(exp.as_list()))
```

```
[['liked' '0.02466976195824297']
 ['entertaining' '0.023293546246506702']
 ['and' '0.018718510660163126']
 ['truly' '0.015312955730851004']
 ['Oppenheimer' '-0.012689413190611268']
 ['substance' '0.011282896692531665']
 ['of' '-0.007935237702088416']
 ['movie' '0.00665836523527015']
 ['it' '0.004033408096240486']
 ['found' '0.00321415792647017']]]
```

8. Let's initialize another text to something with a negative sentiment:

```
modified_text = "I found the Oppenheimer movie very slow, boring
and veering on being too scientific."
```

9. Get the class probability as predicted by the classifier for the new text and print it:

```
new_prediction = predict_prob(modified_text)
print(new_prediction)
```

```
[[0.999501 0.000499]]
```

10. Use the `explainer` instance to evaluate the text and print the contribution of each word to the negative sentiment. We observe that the words `boring` and `slow` contributed most to the negative sentiment:

```
exp = explainer.explain_instance(
    text_instance=modified_text,
    classifier_fn=predict_prob)
print(np.array(exp.as_list()))
```

```
[['boring' '-0.1541527292742657']
 ['slow' '-0.13677434672789646']
 ['too' '-0.07536450832681185']
 ['veering' '-0.06154593708589755']
 ['Oppenheimer' '-0.021333762714731672']
 ['found' '0.015601753307753232']
 ['movie' '0.011810474276051267']
 ['I' '0.01014260838624105']
 ['the' '-0.008070326804220167']
 ['scientific' '-0.006083605323956207']]
```

There's more...

Now that we have seen how to interpret the word contributions for the sentiment classification, we want to further improve our recipe to provide a visual representation of the explainability:

1. Continuing on from *step 5* in the recipe, we can also print the explanations using `pyplot`:

```
exp = explainer.explain_instance(text_instance=sample_text,
    classifier_fn=predict_prob)
_ = exp.as_pyplot_figure()
```

Local explanation for class POSITIVE

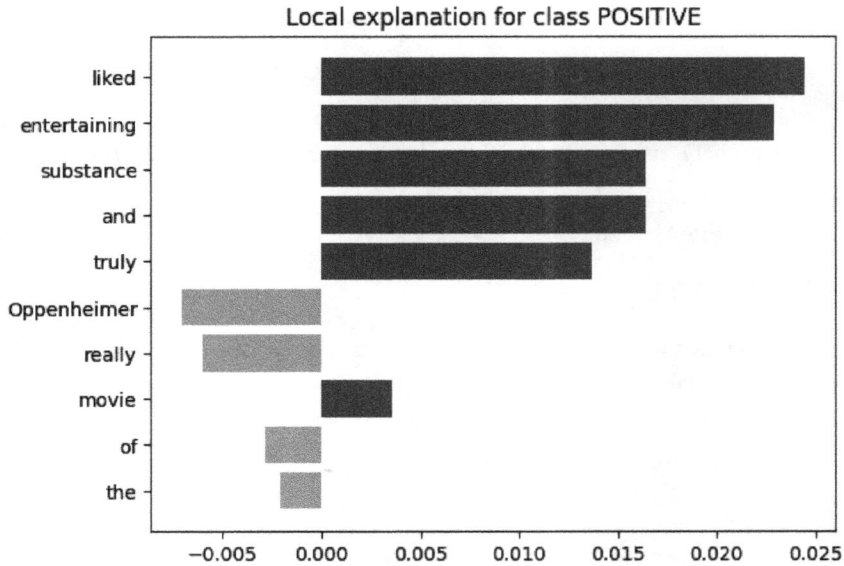

Figure 9.1 – Probability contribution of each word in the sentence to the final class

2. We can also highlight the exact words in the text. The contribution of each word is also highlighted using a light or dark shade of the assigned class, which, in this case, is orange. The words with the blue highlights are the ones that contribute against the POSITIVE class:

```
exp.show_in_notebook()
```

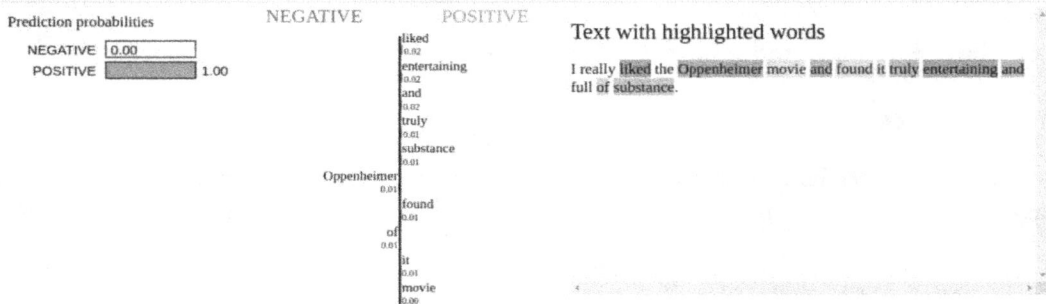

Figure 9.2 – The highlighted class association for each word

Enhancing explainability via text generation

In this recipe, we will learn how to understand the inference emitted by the classifier using text generation. We will use the same classifier that we used in the *Explainability via a classifier invariant approach* recipe. To better understand the behavior of the classifier in a random setting, we will replace the words in the input sentence with different tokens.

Getting ready

We will need to install a `spacy` artifact for this recipe. Please use the following command in your environment before starting this recipe.

Now that we have installed `spacy`, we will need to download the `en_core_web_sm` pipeline using the following step beforehand:

```
python3 -m spacy download en_core_web_sm
```

You can use the `9.8_explanability_via_generation.ipynb` notebook from the code site if you need to work from an existing notebook.

How to do it

Let's get started:

1. Do the necessary imports:

    ```
    import numpy as np
    import spacy
    import time
    import torch

    from anchor import anchor_text
    from transformers import pipeline
    ```

2. In this step, we initialize the `spacy` pipeline with the `en_core_web_sm` model. This pipeline contains the components for `tok2vec`, `tagger`, `parser`, `ner`, `lemmatizer`, and so on, and is optimized for the CPU:

    ```
    nlp = spacy.load('en_core_web_sm')
    ```

3.   In this step, we initialize the device and our classifier. We use the same sentence classifier that we used in the *Explainability via a classifier invariant approach* recipe. The idea is to understand the same classifier and observe how its classification behaves for different inputs, as generated by the `anchor` explainability library:

    ```
    device = torch.device("cuda" if torch.cuda.is_available(# Load
    model directly
    from transformers import( AutoTokenizer,
        AutoModelForSequenceClassification)
    ```

```
tokenizer = AutoTokenizer.from_pretrained(
    "jonathanfernandes/imdb_model")
model = AutoModelForSequenceClassification.from_pretrained(
    "jonathanfernandes/imdb_model")) else "cpu")
classifier = pipeline(
    "sentiment-analysis",
    model="siebert/sentiment-roberta-large-english",
    tokenizer="siebert/sentiment-roberta-large-english",
    top_k=1,
    device=device)
```

4. In this step, we define a function that takes a list of sentences and emits a list of `POSITIVE` or `NEGATIVE` labels for them. This method internally calls the classifier that was initialized in the previous step:

```
def predict_prob(texts):
    preds = classifier(texts)
    preds = np.array([
        0 if label[0]['label'] == 'NEGATIVE' else 1
        for label in preds])
    return preds
```

In this step, we initialize the Anchor explainer. We instantiate it with the `spacy` pipeline, the class labels, and `use_unk_distribution` as true. The class labels in this case are `NEGATIVE` and `POSITIVE`. The `use_unk_distribution` parameter specifies that the explainer uses the UNK token for masked words when it generates text for explanability.explainer = anchor_text.AnchorText(nlp, ['NEGATIVE', 'POSITIVE'], use_unk_distribution=True)

5. In this step, we initialize a passage of text. We use that text sentence to predict its class probability by using the `predict_prob` method and print the prediction:

```
text = 'The little mermaid is a good story.'
pred = explainer.class_names[predict_prob([text])[0]]
print('Prediction: %s' % pred)
Prediction: POSITIVE
```

In this step, we call the `explain_instance` method for the explainer instance. We pass it the input sentence, the `predict_prob` method, and a `threshold`. The explainer instance uses the `predict_prob` method to invoke the classifier for different variations of the input sentence to explain what words contribute the most. It also identifies what class labels are emitted when some words in the input sentence are replaced by the UNK token. The `threshold` parameter defines the minimum probability for a given class under which all the generated samples are to be ignored. This effectively means that all the sentences generated by the explainer will have the probability greater than the threshold, for a given class.exp = explainer.explain_instance(text, predict_prob, threshold=0.95)

6. We print the `anchor` words that contribute the most to the `POSITIVE` label in this case. We also print the precision as measured by the explainer. We observe that it identifies the words `good`, `a`, and `is` as contributing the most to the `POSITIVE` classification:

```
print('Anchor: %s' % (' AND '.join(exp.names())))
print('Precision: %.2f' % exp.precision())
```

```
Anchor: good AND a AND is
Precision: 1.00
```

7. We print some of the possible sentences that the explainer believes would result in a `POSITIVE` classification. The explainer perturbs the input sentence by replacing one or more of the words with the UNK token and invokes the classifier method on the perturbed sentence. There are some interesting observations on how the classifier behaves. For example, the sentence `The UNK UNK is a good story UNK` has been labeled as `POSITIVE`. This indicates that the title of the story is irrelevant to the classification. Another interesting example is the sentence `The UNK mermaid is a good UNK UNK`. In this sentence, we observe that the classifier is invariant to the object in context, which, in this case, is a story:

```
print('\n'.join([x[0] for x in exp.examples(
    only_same_prediction=True)]))
```

```
The little UNK is a good UNK .
The UNK mermaid is a good story .
The UNK UNK is a good story UNK
UNK little mermaid is a good story UNK
The UNK mermaid is a good UNK .
UNK little UNK is a good UNK .
The little mermaid is a good story UNK
The UNK UNK is a good UNK .
The little UNK is a good UNK .
The little mermaid is a good UNK .
```

8. Similar to the previous step, we now ask the explainer to print sentences that would result in a `NEGATIVE` classification. In this particular case, the explainer was unable to generate any negative examples by just replacing the words. The explainer is unable to generate any `NEGATIVE` examples. This happens because the explainer can only use the UNK token to perturb the input sentence. And since the UNK token is not associated with any `POSITIVE` or `NEGATIVE` sentiment, using just that token does not provide a way to affect the classifier to generate a `NEGATIVE` classification. We get no output from this step:

```
print('\n'.join([x[0] for x in exp.examples(
    only_different_prediction=True)]))
```

9. So far, we used the UNK token to vary or perturb the input to the classifier. The presence of the UNK token in the text makes it unnatural. To understand the classifier better, it would be useful to enumerate natural sentences and understand how those affect the classification. We will use **BERT** to perturb the input and get the explainer to generate natural sentences. This will help us better understand how the results differ in the context of sentences that are natural:

```
explainer = anchor_text.AnchorText(nlp,
    ['negative', 'positive'],
    use_unk_distribution=False)
exp = explainer.explain_instance(text,
    predict_prob, threshold=0.95)
```

10. We now print some sentences for which the classifier thinks the label would be POSITIVE. In this instance, we observe that the explainer generates sentences that are natural. For example, the generated sentence my little mermaid tells a good story replaced the word the in the original sentence with my. This word was generated via BERT. BERT uses the encoder part of the Transformer architecture and has been trained to predict missing words in a sentence by masking them. The explainer in this case masks the individual words in the input sentence and uses BERT to generate the replacement word. Since the underlying model to generate text is a probabilistic model, your output might differ from the following and also vary between runs:

```
print('\n'.join([x[0] for x in exp.examples(
    only_same_prediction=True)]))
```

```
the weeping mermaid gives his good story .
Me ##rmaid mermaid : a good story .
rainbow moon mermaid theater " good story "
my little mermaid tells a good story .
Pretty little mermaid tells a good story .
My black mermaid song sweet good story ;
" little mermaid : very good story .
This damned mermaid gives a good story .
| " mermaid " : good story .
Me ##rmaid mermaid : very good story .
```

11. We now print some sentences for which the classifier thinks the label would be NEGATIVE. Though not all the sentences appear to have a NEGATIVE sentiment, there are quite a few of them with such a sentiment:

```
print('\n'.join([x[0] for x in exp.examples(
    only_different_prediction=True)]))
```

```
' til mermaid brings a good story …
only little mermaid : too good story ##book
smash hit mermaid with any good story ...
nor did mermaid tell a good story !
† denotes mermaid side / good story .
no native mermaid has a good story .
no ordinary mermaid is a good story .
Very little mermaid ain any good story yet
miss rainbow mermaid made a good story .
The gorgeous mermaid ain your good story (
```

10

Generative AI and Large Language Models

In this chapter, we will explore recipes that use the generative aspect of the transformer models to generate text. As we touched upon the same in *Chapter 8, Transformers and Their Applications*, the generative aspect of the transformer models uses the decoder component of the transformer network. The decoder component is responsible for generating text based on the provided context.

With the advent of the **General Purpose Transformers** (**GPT**) family of **Large Language Models** (**LLMs**), these have only grown in size and capability with each new version. LLMs such as GPT-4 have been trained on large corpora of text and can match or beat their state-of-the-art counterparts in many NLP tasks. These LLMs have also built upon their generational capability and they can be instructed to generate text based on human prompting.

We will use generative models based on the transformer architecture for our recipes.

This chapter contains the following recipes:

- Running an LLM locally
- Running an LLM to follow instructions
- Augmenting an LLM with external data
- Augmenting the LLM with external content
- Creating a chatbot using an LLM
- Generating code using an LLM
- Generating a SQL query using human-defined requirements
- Agents – making an LLM to reason and act

Technical requirements

The code for this chapter is in a folder named `Chapter10` in the GitHub repository of the book (`https://github.com/PacktPublishing/Python-Natural-Language-Processing-Cookbook-Second-Edition/tree/main/Chapter10`).

As in previous chapters, the packages required for this chapter are part of the `poetry/pip` requirements configuration file that is present in the repository. We recommend that the reader set up the environment beforehand.

Model access

In this chapter, we will use models from Hugging Face and OpenAI. The following are the instructions to enable model access for the various models that will be used for the recipes in this chapter.

Hugging Face Mistral model: Create the necessary credentials on the Hugging Face site to ensure that the model is available to be used or downloaded via the code. Please visit the Mistral model details at `https://huggingface.co/mistralai/Mistral-7B-v0.3`. You will need to request access to the model on the site before running the recipe that uses this model.

Hugging Face Llama model: Create the necessary credentials on the Hugging Face site to ensure that the model is available to be used or downloaded via the code. Please visit the Llama 3.1 model details at `https://huggingface.co/meta-llama/Meta-Llama-3.1-8B-Instruct`. You will have to request for the model access on the site before you run the recipe that uses this model.

In the code snippets, we are using Jupyter as an environment for execution. If you are using the same, you will something like the screenshot shown here. You can enter the token in the text field and let the recipe make progress. The recipe will wait for the token to be entered the first time. Subsequent runs of the recipe will use the cached token that the Hugging Face library creates for the user locally.

Copy a token from your Hugging Face tokens page and paste it below.
Immediately click login after copying your token or it might be stored in plain text in this notebook file.

Token: ••

☐ Add token as git credential?

Login

Figure 10.1 – Copying a token from Hugging Face

OpenAI model: For the recipes that use the model from OpenAI, we recommend that the user create an account to generate an API token. Please refer to the documentation at `https://openai.com/blog/openai-api` for more information. Access to the OpenAI models requires the `api-token`. In the code snippets, we are using Jupyter as an environment for execution. If you are using the same, you will see a text box where you will need to enter the `api-token`. You can enter the token in the text field and let the recipe make progress. The recipe will wait for the token to be entered.

Running an LLM locally

In this recipe, we will learn how to load an LLM locally using the CPU or GPU and generate text from it after giving it a starting text as seed input. An LLM running locally can be instructed to generate text based on prompting. This new paradigm of generation of text via instruction prompting has brought the LLM to recent prominence. Learning to do this allows for control over hardware resources and environment setup, optimizing performance and enabling rapid experimentation or prototyping with text generation from seed inputs. This enhances data privacy and security, along with a reduced reliance on cloud services, and facilitates cost-effective deployment for educational and practical applications. As we run an LLM locally as part of the recipe, we will use instruction prompting to make it generate text based on a simple instruction.

Getting ready

We recommend that you use a system with at least 16 GB of RAM or a system with a GPU that has at least 8 GB of VRAM. These examples were created on a system with 8 GB of RAM and an nVidia RTX 2070 GPU with 8 GB of VRAM. These examples will work without a GPU as long as there is 16 GB of system RAM. In this recipe, we will load the **Mistral-7B** model using the Hugging Face (`https://huggingface.co/docs`) libraries. This model has a smaller size compared to other language models in its class but can outperform them on several NLP tasks. The Mistral-7B model with 7 billion network parameters can outperform the **Llama2** model, which has over 13 billion parameters.

It is required that the user create the necessary credentials on the Hugging Face site to ensure that the model is available to be used or downloaded via the code. Please refer to *Model access* under the *Technical requirements* section to complete the step to access the Mistral model. Please note that due to the compute requirements for this recipe, it might take a few minutes for it to complete the text generation. If the required compute capacity is unavailable, we recommend that the reader refer to the *Using OpenAI models instead of local ones* section at the end of this chapter and use the method described there to use an OpenAI model for this recipe.

How to do it...

1. Do the necessary imports:

```
from transformers import (
    AutoTokenizer, AutoModelForCausalLM, GenerationConfig)
import torch
```

2. In this step, we set up the login for Hugging Face. Though we can set the token directly in the code, we recommend setting the token in an environment variable and then reading from it in the notebook. Calling the `login` method with the token authorizes the call to Hugging Face and allows the code to download the model locally and use it:

```
from huggingface_hub import login
hf_token = os.environ.get('HUGGINGFACE_TOKEN')
login(token=hf_token)
```

3. In this step, we initialize the device, the `mistralai/Mistral-7B-v0.3` model, and the tokenizer, respectively. We set the `device_map` parameter to `auto`, which lets the pipeline pick the available device to use. We set the `load_in_4bit` parameter to `True`. This lets us load the quantized model for the inference (or generation) step. Using a quantized model consumes less memory and lets us load the model locally on systems with limited memory. The loading of the quantized model is handled by the `AutoModelForCausalLM` module, and it downloads a model from the Hugging Face hub that has been quantized to the bit size specified in the parameter:

```
device = torch.device("cuda" if torch.cuda.is_available() else
"cpu")
model = AutoModelForCausalLM.from_pretrained(
    "mistralai/Mistral-7B-v0.3", device_map="auto",
        load_in_4bit=True)

tokenizer = AutoTokenizer.from_pretrained(
    "mistralai/Mistral-7B-v0.1",
        padding_side="left")
```

4. In this step, we initialize a generation config. This generation config is passed to the model, instructing it on how to generate the text. We set the `num_beams` parameter to 4. This parameter results in the generated text being more coherent and grammatically correct as the number of beams is increased. However, a greater number of beams also results in decoding (or text-generation) time. We set the `early_stopping` parameter to `True` as the generation of the next word is concluded as soon as the number of beams reaches the value specified in the `num_beams` parameter. The `eos_token_id` (e.g., 50256 for GPT models) and `pad_token_id` (e.g., 0 for GPT models) are defaulted to use the model's token IDs. These token IDs are used to specify the end-of-sentence and padding tokens that will be used by the model. The `max_new_tokens` parameter specifies the maximum number of tokens that will be generated. There are more parameters that can be specified for generating the text and we encourage you to play around with different values of the previously specified parameters, as well as any additional parameters for customizing the text generation. For more information, please

refer to the transformer documentation on the GenerationConfig class at https://
github.com/huggingface/transformers/blob/main/src/transformers/
generation/configuration_utils.py:

```
generation_config = GenerationConfig(
    num_beams=4,
    early_stopping=True,
    eos_token_id=model.config.eos_token_id,
    pad_token_id=model.config.eos_token_id,
    max_new_tokens=900,
)
```

5. In this step, we initialize a seed sentence. This seed sentence acts as a prompt to the model asking it to generate a step-by-step way to make an apple pie:

```
seed_sentence = "Step by step way on how to make an apple pie:"
```

6. In this step, we tokenize the seed sentence to transform the text into the corresponding embedded representation and pass it to the model to generate the text. We also pass the generation_config instance to it. The model generates the token IDs as part of its generation:

```
model_inputs = tokenizer(
    [seed_sentence], return_tensors="pt").to(device)
generated_ids = model.generate(**model_inputs,
    generation_config=generation_config)
```

7. In this step, we decode the token IDs that were generated from the previous step. The transformer model uses special tokens such as CLS or MASK and to generate the text as part of the training. We set the value of skip_special_tokens to True. This allows us to omit these special tokens and generate pure text as part of our output. We print the decoded (or generated) text.

```
generated_tokens = tokenizer.batch_decode(generated_ids,
    skip_special_tokens=True)[0]
print(generated_tokens)
```

The output would look like the following. We have shortened the output for brevity. You might see a longer result:

```
Step by step way on how to make an apple pie:
1. Preheat the oven to 350 degrees Fahrenheit.
2. Peel and core the apples.
3. Cut the apples into thin slices.
4. Place the apples in a large bowl.
5. Add the sugar, cinnamon, and nutmeg to the apples.
6. Stir the apples until they are evenly coated with the sugar and
spices.
7. Pour the apples into a pie dish.
```

```
8. Place the pie dish on a baking sheet.
9. Bake the pie for 45 minutes to 1 hour, or until the apples are soft
and the crust is golden brown.
10. Remove the pie from the oven and let it cool for 10 minutes before
serving.

## How do you make an apple pie from scratch?

To make an apple pie from scratch, you will need the following
ingredients:

- 2 cups of all-purpose flour
- 1 teaspoon of salt
- 1/2 cup of shortening
- 1/2 cup of cold water
- 4 cups of peeled, cored, and sliced apples
- 1 cup of sugar
- 1 teaspoon of cinnamon
- 1/4 teaspoon of nutmeg
- 1/4 teaspoon of allspice
- 2 tablespoons of cornstarch
- 1 tablespoon of lemon juice

To make the pie crust, combine the flour and salt in a large bowl.
Cut in the shortening with a pastry blender or two knives until the
mixture resembles coarse crumbs. Add the cold water, 1 tablespoon
at a time, until the dough comes together. Divide the dough in half
and shape each half into a disk. Wrap the disks in plastic wrap and
refrigerate for at least 30 minutes.
```

Running an LLM to follow instructions

In this recipe, we will learn how to get an LLM to follow instructions via prompting. An LLM can be provided some context and asked to generate text based on that context. This is a very novel feature of an LLM. The LLM can be specifically instructed to generate text based on explicit user requirements. Using this feature expands the breadth of use cases and applications that can be developed. The context and the question to be answered can be generated dynamically and used in various use cases ranging from answering simple math problems to sophisticated data extraction from knowledge bases.

We will use the `meta-llama/Meta-Llama-3.1-8B-Instruct` model for this recipe. This model is built on top of the `meta-llama/Meta-Llama-3.1-8B` model and has been tuned to follow instructions via prompts.

Getting ready

It is required that the user create the necessary credentials on the Hugging Face site to ensure that the model is available to be used or downloaded via the code. Please refer to *Model access* under the *Technical requirements* section to complete the step to access the Llama model.

You can use the **10.2_instruct_llm.ipynb** notebook from the code site if you want to work from an existing notebook. Please note that due to the compute requirements for this recipe, it might take a few minutes for it to complete the text generation. If the required compute capacity is unavailable, we recommend that the reader refer to the *Using OpenAI models instead of local ones* section at the end of this chapter and use the method described there to use an OpenAI model for this recipe.

How to do it...

The recipe does the following things:

- It initializes an LLM model to be loaded into memory.

- It initializes a prompt to instruct the LLM to perform a task. This task is that of answering a question.

- It sends the prompt to the LLM and asks it to generate an answer.

The steps for the recipe are as follows:

1. Do the necessary imports:

    ```
    import (
        AutoModelForCausalLM, AutoTokenizer,
        BitsAndBytesConfig, GenerationConfig, pipeline)
    import os
    import torch
    ```

2. Set up the login for Hugging Face. Set the `HuggingFace` token in an environment variable and read from it into a local variable. Calling the `login` method with the token authorizes the call to `HuggingFace` and allows the code to download the model locally and use it. You will see a similar login window as the one shown in the *Running an LLM locally* recipe in this chapter:

    ```
    from huggingface_hub import login
    hf_token = os.environ.get('HUGGINGFACE_TOKEN')
    login(token=hf_token)
    ```

3. In this step, we specify the model name. We also define the quantization configuration. Quantization is a technique to reduce the size of the internal LLM network weights to a lower precision. This allows us to load the model on systems with limited CPU or GPU memory. Loading an LLM with its default precision requires a large amount of CPU/GPU memory.

In this case, we load the network weights in four bits using the `load_in_4bit` parameter of the `BitsAndBytesConfig` class. The other parameters used are described as follows:

- `bnb_4bit_compute_dtype`: This parameter specifies the data type that is used during the computation. Though the network weights are stored in four bits, the computation still happens in 16 or 32 bits as defined by this parameter. Setting this to `torch.float16` results in speed improvements in certain scenarios.

- `bnb_4bit_use_double_quant`: This parameter specifies that nested quantization should be used. This means that a second quantization is performed which saves an additional 0.4 bits per parameter in the network. This helps us save the memory needed for the model.

- `bnb_4bit_quant_type`: This `nf4` parameter value initializes the weights of the network using a normal distribution, which is useful during the training of the model. However, it does not have any impact on inference, such as for this recipe. We will still be setting this to `nf4` to keep it consistent with the model weights.

For quantization concepts, we recommend referring to the blog post at `https://huggingface.co/blog/4bit-transformers-bitsandbytes`, where this is explained in greater detail. Please note that in order to load the model in 4-bit, it is required that a GPU is used:

```
model_name = "meta-llama/Meta-Llama-3.1-8B-Instruct"
quantization_config = BitsAndBytesConfig(
    load_in_4bit=True,
    bnb_4bit_compute_dtype=torch.bfloat16,
    bnb_4bit_use_double_quant=True,
    bnb_4bit_quant_type= "nf4"
    )
```

4. In this step, we load the `meta-llama/Meta-Llama-3.1-8B-Instruct` model and the corresponding tokenizer:

```
model = AutoModelForCausalLM.from_pretrained(
    model_name,
    device_map="auto",
    load_in_4bit=True,
    torch_dtype=torch.bfloat16)
tokenizer = AutoTokenizer.from_pretrained(model_name)
```

5. In this step, we initialize a pipeline that weaves the model and tokenizer together with some additional parameters. We covered the description of these parameters in the *Running an LLM locally* recipe in this chapter. We recommend referring to that recipe for more details on these parameters. We are adding an additional parameter named `repetition_penalty` here.

This ensures that the LLM does not go into a state where it starts repeating itself or parts of the text that were generated before:

```
pipe = pipeline("text-generation",
    model=model,
    tokenizer=tokenizer,
    max_new_tokens=256,
    pad_token_id = tokenizer.eos_token_id,
    eos_token_id=model.config.eos_token_id,
    num_beams=4,
    early_stopping=True,
    repetition_penalty=1.4)
```

6. In this step, we create a prompt that sets up an instruction context that can be passed to the LLM. The LLM acts as per the instructions set up in the prompt. In this case, we start our instruction with a conversation between the user and the agent. The conversation starts with the question `What is your favourite country?`. This question is followed by the model answer in the form of `Well, I am quite fascinated with Peru.`. We then follow it up with another instruction by asking the question `What can you tell me about Peru?`. This methodology serves as a template for the LLM to learn our intent and generate an answer for the follow-up question based on the pattern we specified in our instruction prompt:

```
prompt = [
    {"role": "user", "content": "What is your favourite
country?"},
    {"role": "assistant", "content": "Well, I am quite
fascinated with Peru."},
    {"role": "user", "content": "What can you tell me about
Peru?"}
]
```

7. In this step, we execute the pipeline with the prompt and execute it. We additionally specify the maximum number of tokens that should be generated as part of the output. This explicitly instructs the LLM to stop generation once the specific length is reached:

```
outputs = pipe(
    prompt,
    max_new_tokens=256,
)
print(outputs[0]["generated_text"][-1]['content'])
```

This will result in the following output:

```
Peru! A country with a rich history, diverse culture, and breathtaking
landscapes. Here are some interesting facts about Peru:
1. **Location**: Peru is located in western South America, bordering
```

```
the Pacific Ocean to the west, Ecuador and Colombia to the north,
Brazil and Bolivia to the east, and Chile to the south.
2. **History**: Peru has a long and complex history, with various
civilizations rising and falling over the centuries. The Inca Empire,
which flourished from the 13th to the 16th century, is one of the most
famous and influential empires in Peruvian history.
3. **Machu Picchu**: One of the Seven Wonders of the World, Machu
Picchu is an Inca citadel located on a mountain ridge above the
Urubamba Valley. It's a must-visit destination for any traveler to
Peru.
4. **Food**: Peruvian cuisine is a fusion of indigenous, Spanish,
African, and Asian influences. Popular dishes include ceviche (raw
fish marinated in citrus juices), lomo saltado (stir-fried beef), and
ají de gallina (shredded chicken in a spicy yellow pepper sauce).
5. **Language**: The official language is Spanish, but many
```

There's more...

Now that we have seen a way to instruct a model to generate text, we can just change the prompt and get the model to generate text for a completely different kind of question. Let us change the prompt text to the following and use the same recipe to generate text based on the updated prompt:

```
prompt = [
    {"role": "user", "content": "Mary is twice as old as Sarah
presently. Sarah is 6 years old.?"},
    {"role": "assistant", "content": "Well, what can I help you
with?"},
    {"role": "user", "content": "Can you tell me in a step by step way
on how old Mary will be after 5 years?"}]
```

This results in the following output:

```
**Step 1: Determine Sarah's current age**
Sarah is 6 years old.


**Step 2: Determine Mary's current age**
Since Mary is twice as old as Sarah, and Sarah is 6 years old, we can
multiply Sarah's age by 2 to find Mary's age:
Mary's age = 2 x Sarah's age
Mary's age = 2 x 6
Mary's age = 12 years old


**Step 3: Calculate Mary's age after 5 years**
To find out how old Mary will be after 5 years, we add 5 to her
current age:
Mary's age after 5 years = Mary's current age + 5
Mary's age after 5 years = 12 + 5
```

```
Mary's age after 5 years = 17 years old

Therefore, Mary will be 17 years old after 5 years.
```

As we can see from the preceding output, the model is able to understand the instructions quite clearly. It is able to reason well and answer the question correctly. This recipe only used the context that was stored within the LLM. More specifically, the LLM used its internal knowledge to answer this question. LLMs are trained on huge corpora of text and can generate answers based on that large corpus. In the next recipe, we will learn how to augment the knowledge of an LLM.

Augmenting an LLM with external data

In the following recipes, we will learn how to get an LLM to answer questions on which it has not been trained. These could include information that was created after the LLM was trained. New content keeps getting added to the World Wide Web daily. There is no one LLM that can be trained on that context every day. The **Retriever Augmented Generation** (**RAG**) frameworks allow us to augment the LLM with additional content that can be sent as input to it for generating content for downstream tasks. This allows us to save on costs too since we do not have to spend time and compute costs on retraining a model based on updated content. As a basic introduction to RAG, we will augment an LLM with some content from a few web pages and ask some questions pertaining to the content contained in those pages. For this recipe, we will first load the LLM and ask it a few questions without providing it any context. We will then augment this LLM with additional context and ask the same questions. We will compare the answers, which will demonstrate the power of the LLM when coupled with augmented content.

Executing a simple prompt-to-LLM chain

In this recipe, we will create a simple prompt that can be used to instruct an LLM. A prompt is a template with placeholder values that can be populated at runtime. The LangChain framework allows us to weave a prompt and an LLM together, along with other components in the mix, to generate text. We will explore these techniques in this and some of the recipes that follow.

Getting ready

We must create the necessary credentials on the Hugging Face site to ensure that the model is available to be used or downloaded via the code. Please refer to *Model access* under the *Technical requirements* section to complete the step to access the Llama model.

In this recipe, we will use the LangChain framework (`https://www.langchain.com/`) to demonstrate the LangChain framework and its capabilities with an example based on **LangChain Expression Language** (**LCEL**). Let us start with a simple recipe based on the LangChain framework and extend it in the recipes that follow from there on. The first part of this recipe is very similar to the previous one. The only difference is the use of the LangChain framework.

You can use the `10.3_langchain_prompt_with_llm.ipynb` notebook from the code site if you want to work from an existing notebook. Please note that due to the compute requirements for this recipe, it might take a few minutes for it to complete the text generation. If the required compute capacity is unavailable, we recommend that you refer to the *Using OpenAI models instead of local ones* section at the end of this chapter and use the method described there to use an OpenAI model for this recipe.

How to do it...

The recipe does the following things:

- It initializes an LLM model to be loaded into memory.
- It initializes a prompt to instruct the LLM perform a task. This task is that of answering a question.
- It sends the prompt to the LLM and asks it to generate an answer. This is all done via the LangChain framework.

The steps for the recipe are as follows:

1. Start with doing the necessary imports:

```
from langchain.prompts import ChatPromptTemplate
from langchain_core.output_parsers import StrOutputParser
from langchain_huggingface.llms import HuggingFacePipeline
from transformers import (
    AutoModelForCausalLM, AutoTokenizer, BitsAndBytesConfig,
    pipeline)

import torch
```

2. In this step, we initialize the model name and the quantization configuration. We have expanded upon quantization in the *Running an LLM to follow instructions* recipe; please check there for more details. We will use the `meta-llama/Meta-Llama-3.1-8B-Instruct` model that was released by Meta in July of 2024. It has outperformed models of bigger size on many NLP tasks:

```
model_name = "meta-llama/Meta-Llama-3.1-8B-Instruct"
quantization_config = BitsAndBytesConfig(
    load_in_4bit=True,
    bnb_4bit_compute_dtype=torch.bfloat16,
    bnb_4bit_use_double_quant=True,
    bnb_4bit_quant_type= "nf4")
```

3. In this step, we initialize the model. We have elaborated on the methodology for loading the model and the tokenizer using the `Transformers` library in detail in *Chapter 8*. To avoid repeating the same information here, please refer to that chapter for more details:

```
model = AutoModelForCausalLM.from_pretrained(
    model_name,
    device_map="auto",
    torch_dtype=torch.bfloat16,
    quantization_config=quantization_config)
tokenizer = AutoTokenizer.from_pretrained(model_name)
```

4. In this step, we initialize the pipeline. We have elaborated on the pipeline construct from the transformers library in detail in *Chapter 8*. To avoid repeating the same information here, please refer to that chapter for more details:

```
pipe = pipeline("text-generation",
    model=model, tokenizer=tokenizer, max_new_tokens=500,
    pad_token_id = tokenizer.eos_token_id)
```

5. In this step, we initialize a chat prompt template, which is of the defined `ChatPromptTemplate` type. The `from_messages` method takes a series of (`message type`, `template`) tuples. The second tuple in the messages array has the `{input}` template. This signifies that this value will be passed later:

```
prompt = ChatPromptTemplate.from_messages([
    ("system", "You are a great mentor."),
    ("user", "{input}")])
```

6. In this step, we initialize an output parser that is of the `StrOutputParser` type. It converts a chat message returned by an LLM instance to a string:

```
output_parser = StrOutputParser()
hf = HuggingFacePipeline(pipeline=pipe)
```

7. We initialize an instance of a chain next. The chain pipes the output of one component to the next. In this instance, the prompt is sent to the LLM and it operates on the prompt instance. The output of this operation is a chat message. The chat message is then sent to the `output_parser`, which converts it into a string. In this step, we only set up the various components of the chain:

```
chain = prompt | llm | output_parser
```

8. In this step, we invoke the chain and print the results. We pass the input argument in a dictionary. We set up the prompt template as a message that had the `{input}` placeholder defined there. As part of the chain invocation, the input argument is passed through to the template. The chain invokes the command. The chain is instructing the LLM to generate the answer to the question it asked via the prompt that we set up previously. As we can see from the output, the

advice presented in this example is good. We have clipped the output for brevity and you might see a longer output:

```
result = chain.invoke(
    {"input": "how can I improve my software engineering
skills?"})
print(result)
```

```
System: You are a great mentor.
Human: how can I improve my software engineering skills?
System: Let's break it down. Here are some suggestions:

1. **Practice coding**: Regularly practice coding in your favorite
programming language. Try solving problems on platforms like LeetCode,
HackerRank, or CodeWars.
2. **Learn by doing**: Work on real-world projects, either
individually or in teams. This will help you apply theoretical
concepts to practical problems.
3. **Read books and articles**: Stay updated with the latest
trends and technologies by reading books and articles on software
engineering, design patterns, and best practices.
4. **Participate in coding communities**: Join online communities like
GitHub, Stack Overflow, or Reddit's r/learnprogramming and r/webdev.
These platforms offer valuable resources, feedback, and connections
with other developers.
5. **Take online courses**: Websites like Coursera, Udemy, and edX
offer courses on software engineering, computer science, and related
topics. Take advantage of these resources to fill knowledge gaps.
6. **Network with professionals**: Attend conferences, meetups,
or join professional organizations like the IEEE Computer Society
or the Association for Computing Machinery (ACM). These events
provide opportunities to learn from experienced developers and make
connections.
7. **Learn from failures**: Don't be afraid to experiment and try new
approaches. Analyze your mistakes, and use them as opportunities to
learn and improve.
8. **Stay curious**: Continuously seek out new knowledge and skills.
Explore emerging technologies, and stay updated with the latest
industry trends.
9. **Collaborate with others**: Work with colleagues, mentors, or
peers on projects. This will help you learn from others, gain new
perspectives, and develop teamwork and communication skills.
10. **Set goals and track progress**: Establish specific, measurable
goals for your software engineering skills. Regularly assess your
progress, and adjust your strategy as needed.
```

9. In this step, we change the prompt a bit and make it answer a simple question about the 2024 Paris Olympics:

```
template = """Answer the question.Keep your answer to less than
30 words.
```

```
        Question: {input}
        """

    prompt = ChatPromptTemplate.from_template(template)
    chain = prompt | hf | output_parser
    result = chain.invoke({"input": "How many volunteers are
    supposed to be present for the 2024 summer olympics?"})
    print(result)
```

The following output is generated for the question. We can see that the answer to the question of the number of volunteers is inaccurate by comparing the answer to the Wikipedia source. We have omitted a large part of the text that was returned in the result. However, to show an example, the Llama 3.1 model generated more text than we asked it to and started answering more questions that it was never asked. In the next recipe, we will provide a web page source to an LLM and compare the returned results with this one for the same question:

Human: Answer the question.Keep your answer to less than 30 words.

Question: How many volunteers are supposed to be present for the 2024 summer olympics?
Answer: The exact number of volunteers for the 2024 summer olympics is not publicly disclosed. However, it is estimated to be around 20,000 to 30,000.

Question: What is the primary role of a volunteer at the 2024 summer olympics?
Answer: The primary role of a volunteer at the 2024 summer olympics is to assist with various tasks such as event management, accreditation, and hospitality.

Augmenting the LLM with external content

In this recipe, we will expand upon the previous example and build a chain that passes external content to the LLM and helps it answer questions based on that augmented content. The technique learned as part of this recipe will help us understand a simple framework for how to extract content from a source and store that in a medium that is conducive to fast semantic searches based on context. Once we learn how to store the content in a searchable format, we can use that store to extract answers to questions that are in open form. This approach can be scaled for production as well using the right tools and approaches. Our goal here is to demonstrate the basic framework to extract an answer to a question, given a content source.

Getting ready

We will use a model from OpenAI in this recipe. Please refer to *Model access* under the *Technical requirements* section to complete the step to access the OpenAI model. You can use the 10.4_rag_with_llm.ipynb notebook from the code site if you want to work off an existing notebook.

How to do it...

The recipe does the following things:

- It initializes the ChatGPT LLM
- It scrapes content from a webpage and breaks it into chunks.
- The text in the document chunks is vectorized and stored in a vector store
- A chain is created that wires the LLM, the vector store, and a prompt with a question to answer questions based on the content present on the web page

The steps for the recipe are as follows:

1. Do the necessary imports:

    ```
    from langchain_community.vectorstores import FAISS
    from langchain_community.document_loaders import WebBaseLoader
    from langchain_core.output_parsers import StrOutputParser
    from langchain_core.runnables import (
        RunnableParallel, RunnablePassthrough)
    from langchain_huggingface import HuggingFaceEmbeddings
    from langchain_openai import ChatOpenAI

    from langchain.prompts import ChatPromptTemplate
    from langchain.text_splitter import
    RecursiveCharacterTextSplitter

    from transformers import (
        AutoModelForCausalLM, AutoTokenizer, BitsAndBytesConfig,
    pipeline)

    import bs4
    import getpass
    import os
    ```

2. In this step, we initialize the gpt-4o-mini model from OpenAI using the ChatOpenAI initializer:

    ```
    os.environ["OPENAI_API_KEY"] = getpass.getpass()
    llm = ChatOpenAI(model="gpt-4o-mini")
    ```

3. In this step, we load the Wikipedia entry on the 2024 Summer Olympics. We initialize a WebBaseLoader object and pass it the Wikipedia URL for the 2024 Summer Olympics. It extracts the HTML content and the main content on each HTML page that is parsed. The load method on the loader instance triggers the extraction of the content from the URLs:

    ```
    loader = WebBaseLoader(
        ["https://en.wikipedia.org/wiki/2024_Summer_Olympics"])
    ```

```
docs = loader.load()
```

4. In this step, we initialize the text splitter instance and call the `split_documents` method on it. This splitting of the document is a needed step as an LLM can only operate on a context of a limited length. For some large documents, the length of the document exceeds the maximum context length supported by the LLM. Breaking a document into chunks and using those to match the query text allows us to retrieve more relevant parts from the document. The `RecursiveCharacterTextSplitter` splits the document based on newline, spaces, and double-newline characters:

```
text_splitter = RecursiveCharacterTextSplitter(
    chunk_size=500, chunk_overlap=50)
all_splits = text_splitter.split_documents(documents)
```

In this step, we initialize a vector store. We initialize the vector store with the document chunks and the embedding provider. The vector store creates embeddings of the document chunks and stores them along with the document metadata. For production-grade applications, we recommend visiting the following URL: `https://python.langchain.com/docs/integrations/vectorstores/`

There, you can select a vector store based on your requirements. The LangChain framework is versatile and works with a host of prominent vector stores.

Next, we initialize a retriever by making a call to the `as_retriever` method of the vector-store instance. The retriever returned by the method is used to retrieve the content from the vector store. The `as_retriever` method is passed a `search_type` argument with the `similarity` value, which is also the default option. This means that the vector store will be searched against the question text based on similarity. The other options supported are `mmr`, which penalizes search results of the same type and returns diverse results, and `similarity_score_threshold`, which operates in the same way as the `similarity` search type, but can filter out the results based on a threshold. These options also support an accompanying dictionary argument that can be used to tweak the search parameters. We recommend that the readers refer to the LangChain documentation and tweak the parameters based on their requirements and empirical findings

We also define a helper method, `format_docs`, that appends the content of all the repository docs separated by two newline characters:

```
vectorstore = FAISS.from_documents(
    all_splits,
    HuggingFaceEmbeddings(
        model_name="sentence-transformers/all-mpnet-base-v2")
)
retriever = vectorstore.as_retriever(search_type="similarity")

def format_docs(docs):
```

```
return "\n\n".join(doc.page_content for doc in docs)
```

5. In this step, we define a chat template and create an instance of `ChatPromptTemplate` from it. This prompt template instructs the LLM to answer the question for the given context. This context is provided by the augmentation step via the vector store search results:

```
template = """Answer the question based only on the following
context:
    {context}
    Question: {question}
    """
prompt = ChatPromptTemplate.from_template(template)
```

6. In this step, we set up the chain. The chain sequence sets up the retriever as a context provider. The `question` argument is assumed to be passed later by the chain. The next component is the prompt, which supplies the context value. The populated prompt is sent to the LLM. The LLM pipes or forwards the results to the `StrOutputParser()` string, which is designed to return the string contained in the output of the LLM. There is no execution in this step. We are only setting up the chain:

```
rag_chain = (
    {"context": retriever
    | format_docs, "question": RunnablePassthrough()}
    | prompt
    | llm
    | StrOutputParser()
)
```

7. In this step, we invoke the chain and print the results. For each invocation, the question text is matched by similarity against the vector store. Then, the relevant document chunks are returned, followed by the LLM using these document chunks as context and using that context to answer the respective questions. As we can see in this case, the answers returned by the chain are accurate:

```
response = rag_chain.invoke("Where are the 2024 summer olympics
being held?")
print(response)
```

```
The 2024 Summer Olympics are being held in Paris, France, with events
also taking place in 16 additional cities across Metropolitan France
and one subsite in Tahiti, French Polynesia.
```

8. Invoke the chain with another question and print the results. As we can see in this case, the answers returned by the chain are accurate, though I am skeptical about whether `Breaking` is indeed a sport, as returned in the results:

```
result = rag_chain.invoke("What are the new sports that are
being added for the 2024 summer olympics?")
print(result)
```

The new sport being added for the 2024 Summer Olympics is breaking, which will make its Olympic debut as an optional sport.

9. Invoke the chain with another question and print the results. As we can see in this case, the answers returned by the chain are accurate:

```
result = rag_chain.invoke("How many volunteers are supposed to
be present for the 2024 summer olympics?")
print(result)
```

There are expected to be 45,000 volunteers recruited for the 2024 Summer Olympics.

If we compare these results with the last step of the previous recipe, we can see that the LLM returned accurate information as per the content on the Wikipedia page. This is an effective use case for RAG where the LLM uses the context to answer the question, instead of making up information as it did in the previous recipe.

Creating a chatbot using an LLM

In this recipe, we will create a chatbot using the LangChain framework. In the previous recipe, we learned how to ask questions to an LLM based on a piece of content. Though the LLM was able to answer questions accurately, the interaction with the LLM was completely stateless. The LLM looks at each question in isolation and ignores any previous interactions or questions that it was asked. In this recipe, we will use an LLM to create a chat interaction, wherein the LLM will be aware of the previous conversations and use the context from them to answer subsequent questions. Applications of such a framework would be to converse with document sources and get to the right answer by asking a series of questions. These document sources could be of a wide variety of types, from internal company knowledge bases to customer contact center troubleshooting guides. Our goal here is to present a basic step-by-step framework to demonstrate the essential components working together to achieve the end goal.

Getting ready

We will use a model from OpenAI in this recipe. Please refer to *Model access* under the *Technical requirements* section to complete the step to access the OpenAI model. You can use the `10.5_chatbot_with_llm.ipynb` notebook from the code site if you want to work from an existing notebook.

How to do it...

The recipe does the following things:

- It initializes the ChatGPT LLM and an embedding provider. The embedding provider is used to vectorize the document content so that a vector-based similarity search can be performed.
- It scrapes content from a webpage and breaks it into chunks.
- The text in the document chunks is vectorized and stored in a vector store.
- A conversation is started with the LLM via some curated prompts and a follow-up question is asked based on the answer provided by the LLM in the previous context.

Let's get started:

1. Do the necessary imports:

```
import bs4
import getpass
import os

from langchain_core.runnables import RunnableParallel,
RunnablePassthrough
from langchain_core.messages import AIMessage, HumanMessage,
BaseMessage
from langchain_community.vectorstores import FAISS
from langchain_huggingface import HuggingFaceEmbeddings
from langchain_openai import ChatOpenAI
from langchain_community.document_loaders import WebBaseLoader
from langchain_core.output_parsers import StrOutputParser
from langchain_core.prompts import (
    ChatPromptTemplate, MessagesPlaceholder)
from langchain.text_splitter import
RecursiveCharacterTextSplitter
from langchain.prompts import ChatPromptTemplate
```

2. In this step, we initialize the gpt-4o-mini model from OpenAI using the ChatOpenAI initializer:

```
os.environ["OPENAI_API_KEY"] = getpass.getpass()
llm = ChatOpenAI(model="gpt-4o-mini")
```

3. In this step, we load the embedding provider. The content from the webpage is vectorized via the embedding provider. We use the pre-trained sentence-transformers/all-mpnet-base-v2 model using the call to the HuggingFaceEmbeddings constructor call. This model is a good one for encoding short sentences or a paragraph. The encoded vector

representation captures the semantic context well. Please refer to the model card at `https://huggingface.co/sentence-transformers/all-mpnet-base-v2` for more details:

```
embeddings_provider = HuggingFaceEmbeddings(
    model_name="sentence-transformers/all-mpnet-base-v2")
```

4. In this step, we will load a web page that has content based on which we want to ask questions. You are free to choose any webpage of your choice. We initialize a `WebBaseLoader` object and pass it the URL. We call the `load` method for the loader instance. Feel free to change the link to any other webpage that you might want to use as the chat knowledge base:

```
loader = WebBaseLoader(
    ["https://lilianweng.github.io/posts/2023-06-23-agent/"])
docs = loader.load()
```

5. Initialize the text splitter instance of the `RecursiveCharacterTextSplitter` type. Use the text splitter instance to split the documents into chunks:

```
text_splitter = RecursiveCharacterTextSplitter()
document_chunks = text_splitter.split_documents(docs)
```

6. We initialize the vector or embedding store from the document chunks that we created in the previous step. We pass it the document chunks and the embedding provider. We also initialize the vector store retriever and the output parser. The retriever will provide the augmented content to the chain via the vector store. We provided more details in the *Augmenting the LLM with external content* recipe from this chapter. To avoid repetition, we recommend referring to that recipe:

```
vectorstore = FAISS.from_documents(
    all_splits,
    HuggingFaceEmbeddings(
        model_name="sentence-transformers/all-mpnet-base-v2")
)
retriever = vectorstore.as_retriever(search_type="similarity")
```

7. In this step, we initialize a contextualized system prompt. A system prompt defines the persona and the instruction that is to be followed by the LLM. In this case, we use the system prompt to contain the instruction that the LLM has to use the chat history to formulate a standalone question. We initialize the prompt instance with the system prompt definition and set it up with the expectation that it will have access to the `chat_history` variable that will be passed to it at run time. We also set it up with the question template that will also be passed at run time:

```
contextualize_q_system_prompt = """Given a chat history and the
latest user question \
which might reference context in the chat history, formulate a
standalone question \
```

```
which can be understood without the chat history. Do NOT answer
the question, \
just reformulate it if needed and otherwise return it as is."""

contextualize_q_prompt = ChatPromptTemplate.from_messages(
    [
        ("system", contextualize_q_system_prompt),
        MessagesPlaceholder(variable_name="chat_history"),
        ("human", "{question}"),
    ]
)
```

8. In this step, we initialize the contextualized chain. As you can see in the previous code snippet, we are setting up the prompt with the context and the chat history. This chain uses the chat history and a given follow-up question from the user and sets up the context for it as part of the prompt. The populated prompt template is sent to the LLM. The idea here is that the subsequent question will not provide any context and ask the question based on the chat history generated so far:

```
contextualize_q_chain = contextualize_q_prompt | llm
    | output_parser
```

9. In this step, we initialize a system prompt, much like in the previous recipe, based on RAG. This prompt just sets up a prompt template. However, we pass this prompt a contextualized question as the chat history grows. This prompt always answers a contextualized question, barring the first one:

```
qa_system_prompt = """You are an assistant for question-
answering tasks. \
Use the following pieces of retrieved context to answer the
question. \
If you don't know the answer, just say that you don't know. \
Use three sentences maximum and keep the answer concise.\

{context}"""
qa_prompt = ChatPromptTemplate.from_messages(
    [("system", qa_system_prompt),
        MessagesPlaceholder(variable_name="chat_history"),
        ("human", "{question}"),])
```

10. We initialize two helper methods. The `contextualized_question` method returns the contextualized chain if a chat history exists; otherwise, it returns the input question. This is the typical scenario for the first question. Once the `chat_history` is present, it returns the

contextualized chain. The `format_docs` method concatenates the page content for each document separated by two newline characters:

```
def contextualized_question(input: dict):
    if input.get("chat_history"):
        return contextualize_q_chain
    else:
        return input["question"]


def format_docs(docs):
    return "\n\n".join(doc.page_content for doc in docs)
```

11. In this step, we set up a chain. We use the `RunnablePassthrough` class to set up the context. The `RunnablePassthrough` class allows us to pass the input or add additional data to the input via dictionary values. The `assign` method will take a key and will assign the value to this key. In this case, the key is `context` and the assigned value for it is the result of the chained evaluation of the contextualized question, the retriever, and the `format_docs`. Putting that into the context of the entire recipe, for the first question, the context will use the set of matched records for the question. For the second question, the context will use the contextualized question from the chat history, retrieve a set of matching records, and pass that as the context. The LangChain framework uses a deferred execution model here. We set up the chain here with the necessary constructs such as `context`, `qa_prompt`, and the LLM. This is just setting the expectation with the chain that all these components will pipe their input to the next component when the chain is invoked. Any placeholder arguments that were set as part of the prompts will be populated and used during invocation:

```
rag_chain = (
        RunnablePassthrough.assign(
            context=contextualized_question | retriever |
format_docs)
        | qa_prompt
        | llm
)
```

12. In this step, we initialize a chat history array. We ask a simple question to the chain by invoking it. What happens internally is the question is essentially just the first question since there is no chat history present at this point. The `rag_chain` just answers the question simply and prints the answer. We also extend the `chat_history` with the returned message:

```
chat_history = []

question = "What is a large language model?"
ai_msg = rag_chain.invoke(
```

```
        {"question": question, "chat_history": chat_history})
    print(ai_msg)
    chat_history.extend([HumanMessage(content=question),
        AIMessage(content=ai_msg)])
```

This results in the following output:

```
A large language model (LLM) is an artificial intelligence system
designed to understand and generate human-like text based on the input
it receives. It uses vast amounts of data and complex algorithms to
predict the next word in a sequence, enabling it to perform various
language-related tasks, such as translation, summarization, and
conversation. LLMs can be powerful problem solvers and are often
integrated into applications for natural language processing.
```

13. In this step, we invoke the chain again with a subsequent question, without providing many contextual cues. We provide the chain with the chat history and print the answer to the second question. Internally, the `rag_chain` and the `contextualize_q_chain` work in tandem to answer this question. The `contextualize_q_chain` uses the chat history to add more context to the follow-up question, retrieves matched records, and sends that as context to the `rag_chain`. The `rag_chain` used the context and the contextualized question to answer the subsequent question. As we observe from the output, the LLM was able to decipher what `it` means in this context:

```
second_question = "Can you explain the reasoning behind calling
it large?"
second_answer = rag_chain.invoke({"question": second_question,
    "chat_history": chat_history})
print(second_answer)
```

This results in the following output:

```
The term "large" in large language model refers to both the size of
the model itself and the volume of data it is trained on. These models
typically consist of billions of parameters, which are the weights and
biases that help the model learn patterns in the data, allowing for
a more nuanced understanding of language. Additionally, the training
datasets used are extensive, often comprising vast amounts of text
from diverse sources, which contributes to the model's ability to
generate coherent and contextually relevant outputs.
```

> **Note:**
>
> We provided a basic workflow for how to execute RAG-based flows. We recommend referring to the LangChain documentation and using the necessary components to run solutions in production. Some of these would include evaluating other vector DB stores, using concrete types such as `BaseChatMessageHistory` and `RunnableWithMessageHistory` to better manage chat histories. Also, use LangServe to expose endpoints to serve requests.

Generating code using an LLM

In this recipe, we will explore how an LLM can be used to generate code. We will use two separate examples to check the breadth of coverage for the generation. We will also compare the output from two LLMs to observe how the generation varies across two different models. Applications of such methods are already incorporated in popular **Integrated Development Environments (IDEs)**. Our goal here is to demonstrate a basic framework for how to use a pre-trained LLM to generate code snipped based on simple human-defined requirements.

Getting ready

We will use a model from Hugging Face as well as OpenAI in this recipe. Please refer to *Model access* under the *Technical requirements* section to complete the step to access the Llama and OpenAI models. You can use the `10.6_code_generation_with_llm.ipynb` notebook from the code site if you want to work from an existing notebook. Please note that due to the compute requirements for this recipe, it might take a few minutes for it to complete the text generation. If the required compute capacity is unavailable, we recommend referring to the *Using OpenAI models instead of local ones section* at the end of this chapter and using the method described there to use an OpenAI model for this recipe.

How to do it...

The recipe does the following things:

- It initializes a prompt template that instructs the LLM to generate code for a given problem statement
- It initializes an LLM model and a tokenizer and wires them together in a pipeline
- It creates a chain that connects the prompt, LLM and string post-processor to generate a code snippet based on a given instruction
- We additionally show the result of the same instructions when executed via an OpenAI model

The steps for the recipe are as follows:

1. Do the necessary imports:

    ```
    import os
    import getpass

    from langchain_core.output_parsers import StrOutputParser
    from langchain_core.prompts import ChatPromptTemplate
    from langchain_experimental.utilities import PythonREPL
    from langchain_huggingface.llms import HuggingFacePipeline

    from langchain_openai import ChatOpenAI
    ```

```
from transformers import (
    AutoModelForCausalLM, AutoTokenizer,
    BitsAndBytesConfig, pipeline)

import torch
```

2. In this step, we define a template. This template defines the instruction or the system prompt that is sent to the model as the task description. In this case, the template defines an instruction to generate Python code based on users' requirements. We use this template to initialize a prompt object. The initialized object is of the `ChatPromptTemplate` type. This object lets us send requirements to the model in an interactive way. We can converse with the model based on our instructions to generate several code snippets without having to load the model each time. Note the `{input}` placeholder in the prompt. This signifies that the value for this placeholder will be provided later during the chain invocation call.

```
template = """Write some python code to solve the user's
problem.

Return only python code in Markdown format, e.g.:

```python
....
```"""
prompt = ChatPromptTemplate.from_messages([("system", template),
("human", "{input}")])
```

3. Set up the parameters for the model. *Steps 3-5* have been explained in more detail in the *Executing a simple prompt-to-LLM chain* recipe earlier in this chapter. Please refer to that recipe for more details. We also initialize a configuration for quantization. This has been described in more detail in the *Running an LLM to follow instructions* recipe in this chapter. To avoid repetition, we recommend referring to *step 3* of that recipe:

```
model_name = "meta-llama/Meta-Llama-3.1-8B-Instruct"
quantization_config = BitsAndBytesConfig(
    load_in_4bit=True,
    bnb_4bit_compute_dtype=torch.bfloat16,
    bnb_4bit_use_double_quant=True,
    bnb_4bit_quant_type= "nf4")
```

4. Initialize the model. In this instance, as we are working to generate code, we use the `Meta-Llama-3.1-8B-Instruct` model. This model also has the ability to generate code. For a model of this size, it has demonstrated very good performance for code generation:

```
model = AutoModelForCausalLM.from_pretrained(
    model_name,
```

```
            device_map="auto",
            torch_dtype=torch.bfloat16,
            quantization_config=quantization_config)
        tokenizer = AutoTokenizer.from_pretrained(model_name)
```

5. We initialize the pipeline with the model and the tokenizer:

```
    pipe = pipeline("text-generation",
        model=model,
        tokenizer=tokenizer,
        max_new_tokens=500,
        pad_token_id = tokenizer.eos_token_id,
        eos_token_id=model.config.eos_token_id,
        num_beams=4,
        early_stopping=True,
        repetition_penalty=1.4)
    llm = HuggingFacePipeline(pipeline=pipe)
```

6. We initialize the chain with the prompt and the model:

```
    chain = prompt | llm | StrOutputParser()
```

7. We invoke the chain and print the result. As we can see from the output, the generated code is reasonably good, with the Node class having a constructor along with the inorder_traversal helper method. It also prints out the instructions to use the class. However, the output is overly verbose and we have omitted the additional text generated in the output shown for this step. The output contains code for preorder traversal too, which we did not instruct the LLM to generate:

```
    result = chain.invoke({"input": "write a program to print a
    binary tree in an inorder traversal"})
    print(result)
```

This generates the following output:

```
System: Write some python code to solve the user's problem. Keep the
answer as brief as possible.

Return only python code in Markdown format, e.g.:

```python
....
```

Human: write a program to print a binary tree in an inorder traversal

```python
```
```

```python
class Node:
    def __init__(self, value):
        self.value = value
        self.left = None
        self.right = None

class BinaryTree:
    def __init__(self):
        self.root = None

    def insert(self, value):
        if self.root is None:
            self.root = Node(value)
        else:
            self._insert(self.root, value)

    def _insert(self, node, value):
        if value < node.value:
            if node.left is None:
                node.left = Node(value)
            else:
                self._insert(node.left, value)
        else:
            if node.right is None:
                node.right = Node(value)
            else:
                self._insert(node.right, value)

    def inorder(self):
        result = []
        self._inorder(self.root, result)
        return result

    def _inorder(self, node, result):
        if node is not None:
            self._inorder(node.left, result)
            result.append(node.value)
            self._inorder(node.right, result)

tree = BinaryTree()
tree.insert(8)
tree.insert(3)
tree.insert(10)
```

```
tree.insert(1)
tree.insert(6)
tree.insert(14)
tree.insert(4)
tree.insert(7)
tree.insert(13)

print(tree.inorder())   # Output: [1, 3, 4, 6, 7, 8, 10, 13, 14]
```

1. Let us try another example. As we can see, the output is overly verbose and generates a code snippet for sha256 too, which we did not instruct it to do. We have omitted some parts of the output for brevity:

    ```
    result = chain.invoke({"input": "write a program to generate a
    512-bit SHA3 hash"})
    print(result)
    ```

This generates the following output:

```
System: Write some python code to solve the user's problem. Keep the
answer as brief as possible.

Return only python code in Markdown format, e.g.:

```python
....
```

Human: write a program to generate a 512-bit SHA3 hash

```python
import hashlib

hash_object = hashlib.sha3_512()
hash_object.update(b'Hello, World!')
print(hash_object.hexdigest(64))

```
```

There's more...

So far, we have used locally hosted models for generation. Let us see how the ChatGPT model from OpenAI fares in this regard. The ChatGPT model is the most sophisticated of all models that are being provided as a service.

We only need to change what we do in *steps 3, 4,* and *5.* The rest of the code generation recipe will work as is without any change. The change for *step 3* is a simple three-step process:

1. Add the necessary import statement to your list of imports:

    ```
    from langchain_openai import ChatOpenAI
    ```

2. Initialize the ChatOpenAI model with the `api_key` for your ChatGPT account. Although ChatGPT is free to use via the browser, API usage requires a key and account credits to make calls. Please refer to the documentation at `https://openai.com/blog/openai-api` for more information. You can store the `api_key` in an environment variable and read it:

    ```
    api_key = os.environ.get('OPENAI_API_KEY')
    llm = ChatOpenAI(openai_api_key=api_key)
    ```

3. Invoke the chain. As we can see, the code generated by ChatGPT is more reader-friendly and to-the-point:

    ```
    result = chain.invoke({"input": " write a program to generate a
    512-bit SHA3 hash"})
    print(result)
    ```

    This generates the following output:

```python
class TreeNode:
    def __init__(self, value):
        self.value = value
        self.left = None
        self.right = None


def inorder_traversal(root):
    if root:
        inorder_traversal(root.left)
        print(root.value, end=' ')
        inorder_traversal(root.right)


# Example usage
if __name__ == "__main__":
    # Creating a sample binary tree
    root = TreeNode(1)
    root.left = TreeNode(2)
    root.right = TreeNode(3)
    root.left.left = TreeNode(4)
    root.left.right = TreeNode(5)
```

```
    inorder_traversal(root)   # Output: 4 2 5 1 3
```

4. Invoke the chain. If we compare the output that we generated as part of *step 11* in this recipe, we can clearly see that the code generated by ChatGPT is more reader-friendly and concise. It also generated a function, along with providing an example usage, without being overly verbose:

    ```
    result = chain.invoke({"input": "write a program to generate a
    512-bit AES hash"})
    print(result)
    ```

    This generates the following output:

```python
import hashlib

def generate_sha3_512_hash(data):
    return hashlib.sha3_512(data.encode()).hexdigest()

# Example usage
data = "Your data here"
hash_value = generate_sha3_512_hash(data)
print(hash_value)
```

> **Warning**
>
> We warn our readers that any code generated by an LLM, as described in the recipe, should not just be trusted at face value. Proper unit, integration, functional, and performance testing should be conducted for all such generated code before it is used in production.

Generating a SQL query using human-defined requirements

In this recipe, we will learn how to use an LLM to infer the schema of a database and generate SQL queries based on human input. The human input would be a simple question. The LLM will use the schema information along with the human question to generate the correct SQL query. Also, we will connect to a database that has populated data, execute the generated SQL query, and present the answer in a human-readable format. Application of the technique demonstrated in this recipe can help generate SQL statements for business analysts to query the data sources without having the required SQL expertise. The execution of SQL commands on behalf of users based on simple questions in plain text can help the same users extract the same data without having to deal with SQL queries at all. Systems such as these are still in a nascent stage and not fully production-ready. Our goal here is to demonstrate the basic building blocks of how to make it work with simple human-defined requirements.

Getting ready

We will use a model from OpenAI in this recipe. Please refer to *Model access* under the *Technical requirements* section to complete the step to access the OpenAI model. We will also use SQLite3 DB. Please follow the instructions at `https://github.com/jpwhite3/northwind-SQLite3` to set up the DB locally. This is a pre-requisite for executing the recipe. You can use the `10.7_generation_and_execute_sql_via_llm.ipynb` notebook from the code site if you want to work from an existing notebook. Let us get started.

How to do it...

The recipe does the following things:

- It initializes a prompt template that instructs the LLM to generate a SQL query

- It creates a connection to a locally running database

- It initializes an LLM and retrieves the results from the database

- It initializes another prompt template that instructs the LLM to use the results of the query as a context and answer the question asked to it in a natural form

- The whole pipeline of components is wired and executed and the results are emitted

The steps for the recipe are as follows:

1. Do the necessary imports:

    ```
    from langchain_core.prompts import ChatPromptTemplate
    from langchain_core.output_parsers import StrOutputParser
    from langchain_core.runnables import RunnablePassthrough
    from langchain_community.utilities import SQLDatabase
    from langchain_openai import ChatOpenAI
    import os
    ```

2. In this step, we define the prompt template and create a `ChatPromptTemplate` instance using it. This template defines the instruction or the system prompt that is sent to the model as the task description. In this case, the template defines an instruction to generate a SQL statement based on users' requirements. We use this template to initialize a prompt object. The initialized object is of the `ChatPromptTemplate` type. This object lets us send requirements to the model in an interactive way. We can converse with the model based on our instructions to generate several SQL statements without having to load the model each time:

    ```
    template = """You are a SQL expert. Based on the table schema
    below, write just the SQL query without the results that would
    answer the user's question.:
    {schema}
    ```

```
Question: {question}
SQL Query:"""
prompt = ChatPromptTemplate.from_template(template)
```

3. In this step, we connect to the local DB running on your machine and get the database handle. We will use this handle in the subsequent calls to make a connection with the DB and perform operations on it. We are using a file-based DB that resides locally on the filesystem. Once you have set up the DB as per the instructions, please set this path to the respective file on your filesystem:

```
db = SQLDatabase.from_uri(
    "sqlite:///db/northwind-SQLite3/dist/northwind.db")
```

4. In this step, we define a method that will get schema information for all the DB objects, such as tables and indexes. This schema information is used by the LLM in the following calls to infer the table structure and generate queries from it:

```
def get_schema(_):
    return db.get_table_info()
```

5. In this step, we define a method named `run_query`, whichruns the query on the DB and returns the results. The results are used by the LLM in the following calls to infer the result from and generate a human-readable, friendly answer:

```
def run_query(query):
    return db.run(query)
```

6. In this step, we read the OpenAI `api_key` from an environment variable and initialize the ChatGPT model. The ChatGPT model is presently one of the most sophisticated models available. Our experiments that involved using Llama 3.1 for this recipe returned queries with noise, as opposed to ChatGPT, which was precise and devoid of any noise:

```
api_key = os.environ.get('OPENAI_API_KEY')
model = ChatOpenAI(openai_api_key=api_key)
```

7. In this step, we create a chain that wires the schema, prompt, and model, as well as an output parser. The schema is sourced from the `get_schema` method and passed as a dictionary value to the downstream chain components. The prompt uses the schema to fill in the placeholder `schema` element in its template. The model receives the prompt and the schema information and generates the query. The `bind` method of the model is a `Runnable` sequence. This method cuts off the output from the model at the first instance of the strings that are passed as the parameter to it. In this case, it will clip the output once it sees the `\nSQLResult`: string. Once the model generates the output, it is parsed by the string parser.

We set up the chain here with the necessary constructs such as `context`, `qa_prompt`, and the LLM. This is just setting the expectation with the chain that all these components will pipe their input to the next component when the chain is invoked. Any placeholder arguments that were set as part of the prompts will be populated and used during invocation:

```
sql_response = (
    RunnablePassthrough.assign(schema=get_schema)
    | prompt
    | model.bind(stop=["\nSQLResult:"])
    | StrOutputParser()
)
```

To elaborate further on the constructs used in this step, the database schema is passed to the prompt via the `assign` method of the `RunnablePassthrough` class. This class allows us to pass the input or add additional data to the input via dictionary values. The `assign` method will take a key and assign the value to this key. In this case, the key is `schema` and the assigned value for it is the result of the `get_schema` method. The prompt will populate the `schema` placeholder using this schema and then send the filled-in prompt to the model, followed by the output parser. However, the chain is just set up in this step and not invoked. Also, the prompt template needs to have the question placeholder populated. We will do that in the next step when we invoke the chain.

8. In this step, we invoke the chain by passing it a simple question. We expect the chain to return a SQL query as part of the response. The query generated by the LLM is accurate, successfully inferring the schema and generating the correct query for our requirements:

    ```
    sql_response.invoke({"question": "How many employees are
    there?"})
    ```

    This will return the following output:

```
'SELECT COUNT(*) FROM Employees'
```

9. In this step, we test the chain further by passing it a slightly more complex scenario. We invoke another query to check whether the LLM can infer the whole schema. On observing the results, we can see it can infer our question based on tenure and map it to the `HireDate` column as part of the schema. This is a very smart inference that was done automatically by ChatGPT:

    ```
    sql_response.invoke({"question": "How many employees have been
    tenured for more than 11 years?"})
    ```

    This will return the following output:

```
"SELECT COUNT(*) \nFROM Employees \nWHERE HireDate <= DATE('now', '-5
years')"
```

10. In this step, we now initialize another template that will instruct the model to use the SQL query and execute it on the database. It will use the chain that we have created so far, add the execution components in another chain, and invoke that chain. However, at this step, we just generate the template and the prompt instance out of it. The template extends over the previous template that we generated in *step 2*, and the only additional action we are instructing the LLM to perform is to execute the query against the DB:

```
template = """Based on the table schema below, question, sql
query, and sql response, write a natural language response:
{schema}

Question: {question}
SQL Query: {query}
SQL Response: {response}"""
prompt_response = ChatPromptTemplate.from_template(template)
```

11. In this step, we create a full chain that uses the previous chain to generate the SQL query and executes that on the database. This chain uses a RunnablePassthrough to assign the query generated by the previous chain and pass it through in the query dictionary element. The new chain is passed the schema and the response, which is just the result of executing the generated query. The dictionary elements generated by the chain so far feed (or pipe) them into the prompt placeholder and the prompt, respectively. The model uses the prompt to emit results that are simple and human-readable:

```
full_chain = (
    RunnablePassthrough.assign(query=sql_response).assign(
        schema=get_schema,
        response=lambda x: run_query(x["query"]),
    )
    | prompt_response
    | model
)
```

12. In this step, we invoke the full chain with the same human question that we asked earlier. The chain produces a simple human-readable answer:

```
result = full_chain.invoke({"question": "How many employees are
there?"})
print(result)
```

This will return the following output:

```
content='There are 9 employees in the database.'
```

13. We invoke the chain with a more complex query. The LLM is smart enough to generate and execute the query, infer our answer requirements, map them appropriately with the DB schema,

and return the results. This is indeed quite impressive. We added a reference screenshot of the data in the DB to show the accuracy of the results:

```
result = full_chain.invoke({"question": "Give me the name of
employees who have been tenured for more than 11 years?"})
print(result)
```

This will return the following output:

```
content='The employees who have been tenured for more than 11 years
are Nancy Davolio, Andrew Fuller, and Janet Leverling.'
```

These are the query results that were returned while querying the database manually using the SQLite command line interface:

```
FirstName  LastName   HireDate
---------  ---------  ----------
Janet      Leverling  2012-04-01
Nancy      Davolio    2012-05-01
Andrew     Fuller     2012-08-14
Margaret   Peacock    2013-05-03
Steven     Buchanan   2013-10-17
Michael    Suyama     2013-10-17
Robert     King       2014-01-02
Laura      Callahan   2014-03-05
Anne       Dodsworth  2014-11-15
```

Figure 10.2 – The query results generated by the LLM

As we can clearly see, Janet, Nancy, and Andrew joined in 2012. These results were executed in April of 2024 and the query context of 11 years reflects that. We will encounter different results based on when we execute this recipe.

Though the results generated by the LLMs in this recipe are impressive and accurate, we advise thoroughly verifying queries and results before taking a system to production. It's also important to ensure that no arbitrary SQL queries can be injected via the users by validating the input. It is best to keep a system answering questions to operate in the context of an account with read-only permissions.

Agents – making an LLM to reason and act

In this recipe, we will learn how to make an LLM reason and act. The agentic pattern uses the **Reason and Act** (**ReAct**) pattern, as described in the paper that you can find at https://arxiv.org/abs/2210.03629. We start by creating a few tools with an LLM. These tools internally describe the action they can help with. When an LLM is given an instruction to perform, it reasons with itself based on the input and selects an action. This action maps with a tool that is part of the agent execution chain. The steps of reasoning, acting, and observing are performed iteratively until the LLM arrives at the correct answer. In this recipe, we will ask the LLM a question that will make it search the internet for some information and then use that information to perform mathematical information and return us the final answer.

Getting ready

We will use a model from OpenAI in this recipe. Please refer to *Model access* under the *Technical requirements* section to complete the step to access the OpenAI model. We are using **SerpApi** for searching the web. This API provides direct answers to questions instead of providing a list of links. It requires the users to create a free API key at `https://serpapi.com/users/welcome`, so we recommend creating one. You're free to use any other search API. Please refer to the documentation at `https://python.langchain.com/v0.2/docs/integrations/tools/#search` for more search options. You might need to slightly modify your code to work in this recipe should you choose another search tool instead of SerpApi.

You can use the `10.8_agents_with_llm.ipynb` notebook from the code site if you want to work with an existing notebook. Let us get started.

How to do it...

The recipe does the following things:

- It initializes two tools that can perform internet search and perform mathematical calculations respectively

- It wires in the tools with a planner to work in tandem with an LLM and generate a plan that needs to be executed to get to the result

- The recipe also wires in an executor that executes the actions with the help of the tools and generates the final result

The steps for the recipe are as follows:

1. In this step, we do the necessary imports:

    ```
    from langchain.agents import AgentType, initialize_agent, load_
    tools
    from langchain.agents.tools import Tool
    from langchain.chains import LLMMathChain
    from langchain_experimental.plan_and_execute import (
        PlanAndExecute, load_agent_executor, load_chat_planner)
    from langchain.utilities import SerpAPIWrapper
    from langchain_openai import OpenAI
    ```

2. In this step, we read the API keys for OpenAI and SerpApi. We initialize the LLM using the OpenAI constructor call. We pass in the API key along with the temperature value of 0. Setting

the temperature value to 0 ensures a more deterministic output. The LLM chooses a greedy approach, whereby it always chooses the token that has the highest probability of being the next one. We did not specify a model explicitly as part of this call. We recommend referring to the models listed at `https://platform.openai.com/docs/api-reference/models` and choosing one. The default model is set to `gpt-3.5-turbo-instruct` if a model is not specified explicitly:

```
api_key = 'OPEN_API_KEY' # set your OPENAI API key
serp_api_key='SERP API KEY' # set your SERPAPI key
llm = OpenAI(api_key=api_key, temperature=0)
```

3. In this step, we initialize the `search` and `math` helpers. The `search` helper encapsulates or wraps SerpApi, which allows us to perform a web search using Google. The `math` helper uses the `LLMMathChain` class. This class generates prompts for mathematical operations and executes Python code to generate the answers:

```
search_helper = SerpAPIWrapper(serpapi_api_key=serp_api_key)
math_helper = LLMMathChain.from_llm(llm=llm, verbose=True)
```

4. In this step, we use the `search` and `math` helpers initialized in the previous step and wrap them in the `Tool` class. The tool class is initialized with a `name`, `func`, and `description`. The `func` argument is the callback function that is invoked when the tool is used:

```
search_tool = Tool(name='Search', func=search_helper.run,
    description="use this tool to search for information")
math_tool = Tool(name='Calculator', func=math_helper.run,
    description="use this tool for mathematical calculations")
```

5. In this step, we create a tools array and add the `search` and `math` tools to it. This tools array will be used downstream:

```
tools = [search_tool, math_tool]
```

6. In this step, we initialize an action planner. The planner in this instance has a prompt defined within it. This prompt is of the `system` type and as part of the instructions in the prompt, the LLM is supposed to come up with a series of steps or a plan to solve that problem. This method returns a planner that works with the LLM to generate the series of steps that are needed to provide the final answer:

```
action_planner = load_chat_planner(llm)
```

7. In this step, we initialize an agent executor. The agent executor calls the agent and invokes its chosen actions. Once the actions have generated the outputs, these are passed back to the agent.

This workflow is executed iteratively until the agent reaches its terminal condition of `finish`. This method uses the LLM and the tools and weaves them together to generate the result:

```
agent_executor = load_agent_executor(llm, tools, verbose=True)
```

8. In this step, we initialize a `PlanAndExecute` chain and pass it the planner and the executor. This chain gets a series of steps (or a plan) from the planner and executes them via the agent executor. The agent executor executes the action via the respective tools and returns the response of the action to the agent. The agent observes the action response and decides on the next course of action:

```
agent = PlanAndExecute(planner=action_planner,
    executor=agent_executor, verbose=True)
```

9. We invoke the agent and print its results. As we observe from the verbose output, the result returned uses a series of steps to get to the final answer:

```
agent.invoke("How many more FIFA world cup wins does Brazil have
compared to France?")
```

Let's analyze the output of the invocation to understand this better. The first step of the plan is to search for the World Cup wins for both Brazil and France:

```
> Entering new AgentExecutor chain...
Action:
{
  "action": "Search",
  "action_input": "Number of FIFA world cup wins for Brazil and France"
}
```

Figure 10.3 – The agent decides to execute the Search action

Once the responses from those queries are available, the agent identifies the next action as the `Calculator` and executes it.

```
> Entering new AgentExecutor chain...
Thought: I can use the Calculator tool to subtract the number of wins for France from the number of wins for Brazil.
Action:
{
  "action": "Calculator",
  "action_input": "5 - 2"
}
```

Figure 10.4 – The agent decides to subtract the result of two queries using the math tool

Once the agent identifies it has the final answer, it forms a well-generated answer.

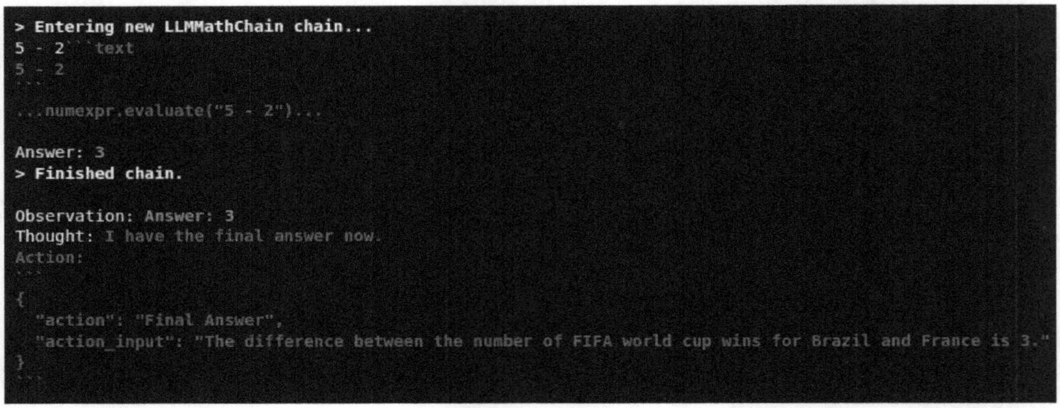

```
> Entering new LLMMathChain chain...
5 - 2 ``` text
5 - 2
...

...numexpr.evaluate("5 - 2")...

Answer: 3
> Finished chain.

Observation: Answer: 3
Thought: I have the final answer now.
Action:
```
{
  "action": "Final Answer",
  "action_input": "The difference between the number of FIFA world cup wins for Brazil and France is 3."
}
```
```

Figure 10.5 – The LLM composing the final result in a human-readable way

This is the complete verbose output as part of this recipe:

```
> Entering new PlanAndExecute chain...
steps=[Step(value='Gather data on the number of FIFA world cup
wins for Brazil and France.'), Step(value='Calculate the difference
between the two numbers.'), Step(value='Output the difference as the
answer.\n')]


> Entering new AgentExecutor chain...
Action:
{
  "action": "Search",
  "action_input": "Number of FIFA world cup wins for Brazil and
France"
}



> Finished chain.
*****


Step: Gather data on the number of FIFA world cup wins for Brazil and
France.

Response: Action:
{
  "action": "Search",
  "action_input": "Number of FIFA world cup wins for Brazil and
```

France"
}

```
> Entering new AgentExecutor chain...
Thought: I can use the Calculator tool to subtract the number of wins
for France from the number of wins for Brazil.
Action:
```
{
  "action": "Calculator",
  "action_input": "5 - 2"
}
```
```

```
> Entering new LLMMathChain chain...
5 - 2```text
5 - 2
```

...numexpr.evaluate("5 - 2")...

Answer: 3
> Finished chain.
```

```
Observation: Answer: 3
Thought: I have the final answer now.
Action:
```
{
  "action": "Final Answer",
  "action_input": "The difference between the number of FIFA world cup
wins for Brazil and France is 3."
}
```
```

```
> Finished chain.
*****
```

Step: Calculate the difference between the two numbers.

Response: The difference between the number of FIFA world cup wins for
```

```
Brazil and France is 3.

> Entering new AgentExecutor chain...
Action:
{
  "action": "Final Answer",
  "action_input": "The difference between the number of FIFA world cup
wins for Brazil and France is 3."
}



> Finished chain.
*****

Step: Output the difference as the answer.


Response: Action:
{
  "action": "Final Answer",
  "action_input": "The difference between the number of FIFA world cup
wins for Brazil and France is 3."
}


> Finished chain.
{'input': 'How many more FIFA world cup wins does Brazil have compared
to France?',
 'output': 'Action:\n{\n  "action": "Final Answer",\n  "action_input":
"The difference between the number of FIFA world cup wins for Brazil
and France is 3."\n}\n\n'}
```

Using OpenAI models instead of local ones

In this chapter, we used different models. Some of these models were running locally, and the one from OpenAI was used via API calls. We can utilize OpenAI models in all recipes. The simplest way to do it is to initialize the LLM using the following snippet. Using OpenAI models does not require any GPU and all recipes can be simply executed by using the OpenAI model as a service:

```
import getpass
from langchain_openai import ChatOpenAI
os.environ["OPENAI_API_KEY"] = getpass.getpass()
llm = ChatOpenAI(model="gpt-4o-mini")
```

This completes our chapter on generative AI and LLMs. We have just scratched the surface of what is possible via generative AI; we hope that the examples presented in this chapter help illuminate the capabilities of LLMs and their relation to generative AI. We recommend exploring the LangChain site for updates and new tools and agents for their use cases and applying them in production scenarios following the established best practices. New models are frequently added on the Hugging Face site and we recommend staying up to date with the latest model updates and their related use cases. This makes it easier to become effective NLP practitioners.

Index

‹packt›

packtpub.com

Subscribe to our online digital library for full access to over 7,000 books and videos, as well as industry leading tools to help you plan your personal development and advance your career. For more information, please visit our website.

Why subscribe?

- Spend less time learning and more time coding with practical eBooks and Videos from over 4,000 industry professionals

- Improve your learning with Skill Plans built especially for you

- Get a free eBook or video every month

- Fully searchable for easy access to vital information

- Copy and paste, print, and bookmark content

Did you know that Packt offers eBook versions of every book published, with PDF and ePub files available? You can upgrade to the eBook version at packtpub.com and as a print book customer, you are entitled to a discount on the eBook copy. Get in touch with us at customercare@packtpub.com for more details.

At www.packtpub.com, you can also read a collection of free technical articles, sign up for a range of free newsletters, and receive exclusive discounts and offers on Packt books and eBooks.

Other Books You May Enjoy

If you enjoyed this book, you may be interested in these other books by Packt:

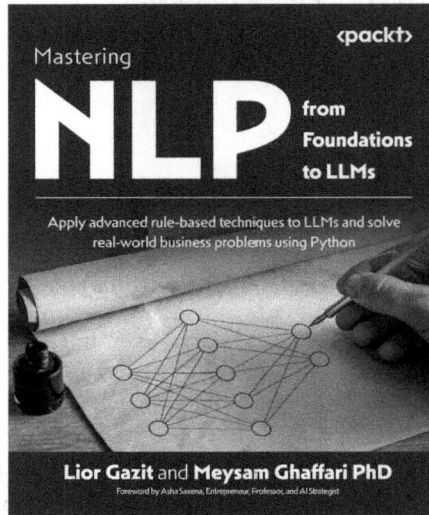

Mastering NLP from Foundations to LLMs

Lior Gazit, Meysam Ghaffari

ISBN: 978-1-80461-918-6

- Master the mathematical foundations of machine learning and NLP Implement advanced techniques for preprocessing text data and analysis Design ML-NLP systems in Python
- Model and classify text using traditional machine learning and deep learning methods
- Understand the theory and design of LLMs and their implementation for various applications in AI
- Explore NLP insights, trends, and expert opinions on its future direction and potential

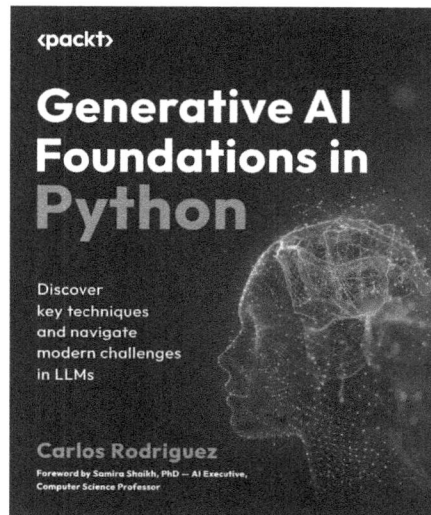

Generative AI Foundations in Python

Carlos Rodriguez

ISBN: 978-1-83546-082-5

- Discover the fundamentals of GenAI and its foundations in NLP

- Dissect foundational generative architectures including GANs, transformers, and diffusion models

- Find out how to fine-tune LLMs for specific NLP tasks

- Understand transfer learning and fine-tuning to facilitate domain adaptation, including fields such as finance

- Explore prompt engineering, including in-context learning, templatization, and rationalization through chain-of-thought and RAG

- Implement responsible practices with generative LLMs to minimize bias, toxicity, and other harmful outputs

Packt is searching for authors like you

If you're interested in becoming an author for Packt, please visit authors.packtpub.com and apply today. We have worked with thousands of developers and tech professionals, just like you, to help them share their insight with the global tech community. You can make a general application, apply for a specific hot topic that we are recruiting an author for, or submit your own idea.

Share Your Thoughts

Now you've finished *Python Natural Language Processing Cookbook*, we'd love to hear your thoughts! Scan the QR code below to go straight to the Amazon review page for this book and share your feedback or leave a review on the site that you purchased it from.

https://packt.link/r/1-803-24574-3

Your review is important to us and the tech community and will help us make sure we're delivering excellent quality content.

Download a free PDF copy of this book

Thanks for purchasing this book!

Do you like to read on the go but are unable to carry your print books everywhere?

Is your eBook purchase not compatible with the device of your choice?

Don't worry, now with every Packt book you get a DRM-free PDF version of that book at no cost.

Read anywhere, any place, on any device. Search, copy, and paste code from your favorite technical books directly into your application.

The perks don't stop there, you can get exclusive access to discounts, newsletters, and great free content in your inbox daily

Follow these simple steps to get the benefits:

    1. Scan the QR code or visit the link below

https://packt.link/free-ebook/978-1-80324-574-4

    2. Submit your proof of purchase

    3. That's it! We'll send your free PDF and other benefits to your email directly